Praise for *Parachuting Cats into Borneo*

"*Parachuting Cats* offers a deep dive into what it takes for our economies and our families to flourish within Earth's finite limits. For all the attention paid to technologies, policies, leadership, and 'corporate social responsibility,' creating the change we want to see in the world means understanding how societies and institutions transform. In the end, it's the system, stupid, that needs transforming. Klimek and AtKisson tell us how to do that. This is a vital read for our turbulent times."—**JOEL MAKOWER**, chairman and executive editor, GreenBiz Group; coauthor of *The New Grand Strategy*

"A fascinating account of the cultural, psychological, and institutional barriers that prevent more change programs from succeeding—and how to overcome them." —**PAUL POLMAN**, CEO, Unilever

"We live in times of continuous accelerating change—as I have personally experienced—and yet we have difficulty adapting to it. That's human nature: We like the comfort of stability and predictability. Here Klimek and AtKisson draw a short and very easy-to-read roadmap for implementing sustainable change. A great effort and recommended reading." —**NANI FALCO BECCALLI**, former president and CEO, GE Europe

"*Parachuting Cats* is a small book with a really big bag of tools for the change agent's toolkit—describing how, when, and where each can best be used. Some are tools for personal change that make one a more effective change agent; others are tools for helping organizations and communities create lasting change. Highly recommend . . . . I could and will reread this book at least ten times and get more out of it each time." —**MAUREEN HART**, executive director, International Society of Sustainability Professionals

"Spanning change management, leadership, strategy, and spirituality, Klimek and AtKisson's volume is an indispensable guide for current and would-be sustainability leaders." —**MICHAEL KOBORI**, vice president of sustainability, Levi Strauss & Co.

"I've been waiting for this book, from these gentlemen, for years. Decades of distilled experience, insight, wisdom, guidance, and delight about engaging the most challenging parts of change—people and groups of people. (Technological innovation is simple by comparison.) Only one in three change initiatives succeed, the authors tell us. This little book, and the thoughtful systems and tools it offers, might just help you boost your odds." —**GIL FRIEND**, chairman and CEO, Natural Logic, Inc.

"The one thing we all have more and more of is CHANGE, and we all need to become more skillful in navigating through it. Klimek and AtKisson are great companions to have with you on your change journey, providing guidance, great stories, and good company." —**PETER HAWKINS**, professor of Leadership, Henley Business School; chairman, Renewal Associates

"This book is a must for anyone who is involved in change processes toward a more equitable, humane, and environmentally friendly world. It is not the usual 'how to do and get what you want' instruction book. No recipes, no safe or proven success guidelines, no software program for making changes happen! It is a book about personal and group *empowerment*. It orients readers to become agents of change based on their own resources and their own creative ideas. And all this for a common purpose: to reach a more sustainable future for all." —**ORTWIN RENN**, scientific director, Institute for Advanced Sustainability Studies

"An apparently endless stream of conferences and workshops is applauding the big transformation toward sustainable development. And is tiring. Real action is not following suit. I see a growing disconnect between advocacy and personal behavior (and the behavior of advocates' home institutions). Yet never before has humankind been in a better position to successfully end hunger and poverty within the limits of ecological boundaries. Never before have there been so many experts and campaigners dedicated to making this planet a better place. But, strangely enough, all this does not yet deliver. Action is often halted.

Advanced thinking is often restricted to special interest groups. Experts are arguing within the boundaries of their own unconnected communities. That is why this book is timely. The authors bridge change attitudes on the personal level and the structural level. They help us understand (and change) the patterns of our very habitudes—and, fortunately, they never forget the importance of changing vested interests and political structures in a democratic society. Absorbing Klimek and AtKisson's recommendations has added value to both my thinking and acting."

—**GÜNTHER BACHMANN**, secretary general, German Council for Sustainable Development; advisor to the Global Network of National Councils for Sustainable Development

"Welcome to the world café—where it's raining, well, cats. Axel Klimek and Alan AtKisson are hosting. Slow down, relax, and prepare to change the way you think about change."

—**JOHN ELKINGTON**, cofounder, Environmental Data Services (ENDS), SustainAbility, and Volans; coauthor of *The Breakthrough Challenge*

"*Parachuting Cats into Borneo* takes change management off the white board and places it into your own hands—inviting you into a café conversation with the authors, who put together a thoughtful collection of practical tools that I found valuable even after 25 years in the sustainability and social change field. Grab a pen and some paper (and a coffee!). This book will take you on a thought journey, best when you have a change process and goal in mind. And who doesn't?" —**GILLIAN MARTIN MEHERS**, managing director, Bright Green Learning; coauthor of *The Climate Change Playbook*

"Many of us need to change ourselves or to bring about change through our work but always get stuck in a rut because we need confirmation to do the right thing. This book helps us enter into conversations to see within and around us and to make that so-needed transformation." —**BIENCE GAWANAS**, former commissioner for Social Affairs, African Union

"*Parachuting Cats into Borneo* is a great guidebook for leaders and individuals who want to create transformational changes in any society, community, organization, workspace, or family they are a part of. The authors have done a great job illuminating not only the most up-to-date 'skills and knowledge' on change processes, such as a system approach and coaching, but also 'attitude and being,' or how leaders can develop themselves and cultivate organizational cultures. I have been using these approaches in Japan and elsewhere in the world, and they have proven to be effective in work for many clients across sectors." —**RIICHIRO ODA**, president and CEO, Change Agent, Inc.

"As a funder, I was drawn to organizations that had both a clear vision for the future and an approach to the inevitable difficulties of change. If this valuable toolkit had been around, I would have sent a copy with every grant check." —**DAVID GRANT,** former president and CEO, Geraldine R. Dodge Foundation; author of *The Social Profit Handbook*

"Change is difficult, and usually takes time, but this book gave me hope that change will happen, whatever time it takes, and guided me through the appropriate sequence of steps I should take to achieve my mission—slowly but steadily. The book presents a combination of concern, determination, and faith: concern about people and nature, the determination to continue the path, and the faith that what we are doing is right." —**BOSHRA SALEM,** director, Office of International Relations, Alexandria University; member, Women in Science Hall of Fame

PARACHUTING CATS *into* BORNEO

PARACHUTING CATS *into* BORNEO

And Other Lessons *from* the Change Café

AXEL KLIMEK

and ALAN ATKISSON

Chelsea Green Publishing

White River Junction, Vermont

Copyright © 2016 by Axel Klimek and Alan AtKisson.
All rights reserved.

Unless otherwise noted, all illustrations are copyright © 2016 by Alan AtKisson.

No part of this book may be transmitted or reproduced in any form
by any means without permission in writing from the publisher.

Editor: Joni Praded
Project Manager: Angela Boyle
Copy Editor: Eileen Clawson
Proofreader: Helen Walden
Indexer: Linda Hallinger
Designer: Melissa Jacobson

Printed in the United States of America.
First printing July, 2016.
10 9 8 7 6 5 4 3 2 1 16 17 18 19 20

Our Commitment to Green Publishing

Chelsea Green sees publishing as a tool for cultural change and ecological stewardship. We strive to align our book manufacturing practices with our editorial mission and to reduce the impact of our business enterprise in the environment. We print our books and catalogs on chlorine-free recycled paper, using vegetable-based inks whenever possible. This book may cost slightly more because it was printed on paper that contains recycled fiber, and we hope you'll agree that it's worth it. Chelsea Green is a member of the Green Press Initiative (www.greenpressinitiative.org), a nonprofit coalition of publishers, manufacturers, and authors working to protect the world's endangered forests and conserve natural resources. *Parachuting Cats into Borneo* was printed on paper supplied by McNaughton & Gunn that contains 100% postconsumer recycled fiber.

Library of Congress Cataloging-in-Publication Data

Names: Klimek, Axel, 1956– author. | AtKisson, Alan, 1960– author.
Title: Parachuting cats into Borneo : and other lessons from the change café
/ Axel Klimek and Alan AtKisson.
Description: White River Junction, Vermont : Chelsea Green Publishing, [2016]
| Includes bibliographical references and index.
Identifiers: LCCN 2016017485| ISBN 9781603586818 (pbk.) |
ISBN 9781603586825 (ebook)
Subjects: LCSH: Organizational change—Management. | Change (Psychology)
Classification: LCC HD58.8 .K579 2016 | DDC 658.4/06—dc23
LC record available at https://lccn.loc.gov/2016017485

Chelsea Green Publishing
85 North Main Street, Suite 120
White River Junction, VT 05001
(802) 295-6300
www.chelseagreen.com

DEDICATION

To our clients:
thank you for putting your trust in us,
and for giving us the opportunity to make
a creative contribution to your important work,
and to learn some lessons worth passing on.

To our colleagues:
thank you for sharing your ideas, and
for listening to our ideas, so that we could put
all of those ideas to work in the world.

To our families:
thank you for your unconditional support
and encouragement as we both traveled out
on our unusual career paths.

And finally, to everyone, everywhere who has
ever been stricken with that impossible-to-ignore
feeling that you must try to make things better:
thank you . . . and we hope that this little book
helps you along the way to success.

CONTENTS

1

Thinking about How We Think about Change

KEY MESSAGES

- Understanding theories and methods for making change is not enough.
- Success in managing change depends just as much on how you perceive and react to the world around you.
- A good companion helps you see things in a new way . . . and build trust in yourself.

When we allow ourselves a moment to reflect on what has happened in the world around us over the past few decades, we can see incredible changes. At the end of 2013, we said farewell to one of the most influential leaders and "change agents" of the last century, Nelson Mandela, who through his vision, personal strength, and principled conviction played a major role in liberating South Africa from apartheid. Although imprisoned for twenty-seven years, he kept his faith that society could be changed in a nonviolent and integrative way. Nelson Mandela's life stands as a monument to the fact that deep change is possible, even against seemingly impossible odds, if there are good leaders who stand up at the right moment. Individuals can lead movements, and movements can change things fundamentally. Fundamental change of this kind is often called "transformation."

Similar moments of proof that transformative change is possible include the dramatic fall of the Berlin Wall in 1989, the nonviolent movement for India's independence led by Mahatma Gandhi, the civil rights movement for black Americans led by Martin Luther King Jr., and the global movement for gender equity led by thousands of women and men across the world (a transformation that is still under way).

Transformation is not always about changing whole societies. The nuclear accident at Fukushima, Japan, in 2011 triggered Germany's *Energiewende*, the country's swift transition away from nuclear and fossil energy and toward renewable energy. The change in Germany had actually been envisioned and planned years earlier; the Fukushima accident proved to be a moment that made it possible to enact those plans.

So even when the media is full of news about what might be called "negative change," we need not look very far to see that positive, transformational change, intended or spontaneous, is successfully happening in the world. Sometimes the reasons for that success can be clearly connected to specific people and events, and sometimes the reasons are not so easy to identify.

On the other hand, we also face another reality regarding change: it doesn't always happen.

As Scott Keller and Carolyn Aiken point out in *The Inconvenient Truth About Change Management*, "In 1996, John Kotter published *Leading Change*. Considered by many to be the seminal work in the field of change management, Kotter's research revealed that only 30 percent of change programs succeed. Since the book's release, literally thousands of books and journal articles have been published on the topic, and courses dedicated to managing change are now part of many major MBA programs. Yet in 2008, a McKinsey survey of 3,199 executives around the world found, as Kotter did, that only one transformation in three succeeds. Other studies over the past ten years reveal remarkably similar results. It seems that, despite prolific output, the field of change management hasn't led to more successful change programs."[1]

We seem to face two different realities. One is that dramatic change is happening around us all the time, often led intentionally by people. The other is that planned organizational change is unlikely to lead to the targeted result.

We assume that if you've chosen to read this book, you are someone who wants to change something, or who has taken on the responsibility of running a change process. If McKinsey's observation is right—and all the thoughts and theories about "change management" have not led to an increase in the success of planned change processes—what can our little book offer you?

We definitely cannot give you the answers and to-do lists that will turn all your change efforts into success stories. The intention of this book is not to provide easy answers; it is to provide a kind of frame for a good conversation among peers or companions. We are fully convinced that all the necessary answers are already there within yourself, or are accessible to you from people and other information sources that are in easy reach. It is fully possible to achieve the change goals that you want (or need) to achieve.

If you don't buy into this general assumption, then it would be much wiser not to start the endeavor at all. Have the courage to say no. The cost involved in every unsuccessful change process is not only that it wastes time and resources but also that it strengthens disbelief in the possibility of change among the people involved.

Starting from our basic assumption that you are the right person at the right time to help change to occur, we believe that the biggest risk to your success lies not in being underinformed about theories and methods; it is in approaching the process with habitual ways of thinking and perceiving, and not reflecting well or deeply enough on your or your organization's belief systems about change.

What would we expect from a good companion, if we were to relate to him or her that we have a challenging task in front of us?

She or he might sit together with us, listen, and ask some deepening questions that help us see hidden patterns—patterns that we didn't see before; patterns in the system that we want to change; patterns in our way of perceiving reality, selecting information, phrasing arguments, and deciding what to do. We would also probably like our companion to help us see and unleash the potential and talents that we are not fully utilizing. From a conversation with a good companion, we would expect a real impact on how we think, feel, and act.

This is our wish for this little book. We wrote it to help you build on what you already know about change; to give you some new ways of seeing things, and naming what you see; to help you be more sensitive and mindful to your own habits and patterns, as well as to the habits and patterns that exist in the system you are trying to impact. At the same time we want to help you focus on your strengths, resources, and experiences and help you—and the people whom you, in turn, will impact—to fully integrate your potential into the work that you do to improve the world.

We can't promise you success. Changing people and systems is always complex, and it includes a large degree of unpredictability. But what we can promise you is an opportunity to deepen your sense of inherent capacity, to see what's happening around you more clearly, to envision the outcome you want more collaboratively, and to respond to the dynamics along the way with greater intelligence and sensitivity.

We believe this will help you get where you want to go.

2

How to Use This Book

KEY MESSAGES

- Welcome to the Change Café, where you are in charge of the conversation.
- We introduce your conversation companions, Axel and Alan.
- There are "four big questions" you need to consider as you approach the process of change.

Just imagine that you are sitting in a nice café with us, Axel and Alan. The three of us have come together to talk about change. You are focused on some change process that you're involved in, and you'd like to create some real, tangible results. We assume that in our conversation, or series of conversations, we would reflect together with you on a wide range of topics—topics that are, from our own experience, crucial to the process of facilitating change.

We believe that you will get the most out of this little book if you can keep that scenario in mind. Into our café conversation we will bring some experience based on years of helping people to design, build, and run complex change processes in companies, communities, and large organizations, such as government agencies. We're focused on spreading the concept and practice of sustainability, so it can become the guiding framework for decision makers and executives and will influence economic systems at the organizational, national, and international level. But what we have learned along the way is applicable to any change process where you, as a change agent, are trying to improve your corner of the world in some way.

As change agents ourselves, we are not limited to the topic of sustainability. We've worked on driving or facilitating change in many other ways, from helping leaders clarify their visions, to improving the efficiency of administrative systems. But sustainability—long-term, systemic health that takes the needs of both people and nature into account—is our passion, and helping to promote sustainability is a responsibility we've chosen. It's a "heart issue" for us.

The more we put our own time and energy into this issue, the more we realize that we have probably picked one of the toughest change topics you can imagine. Why is sustainability so difficult? Because achieving it requires altering some very deeply embedded human habits, concepts, and attitudes. (We will explore this later.) Our focus on "change for sustainability" has thus exposed us to more than our fair share of challenges, in areas ranging from government policy to corporate management and from educational practice to community organizing and more. We've also been fortunate to work with change agents and change programs in dozens of countries around the world, on a very wide range of topics—inside and outside sustainability, but almost always connected to some change of internal culture—and also involving many differences in culture and worldview. We have helped orchestrate change-making efforts for a wide range of clients: from UN programs to US Army bases; from start-up companies with 25 people to multi-national corporations with billions in revenue; from plans to "green" Egypt's economic sectors to helping WWF create strategy for the "blue" (marine) economy. Along the way, we've been taught and mentored by some remarkable people and learned a lot through the endeavors with our clients, and we want to pass on some of the things we've learned.

We feel that our passion and sense of responsibility to promote sustainability doesn't limit us in our ability to be a good companion for you, in thinking about your change initiatives, regardless of the topic that you personally are focused on. Working on sustainability is tough, but it has also been very rewarding, partly because it has exposed us to such a broad range of specific subjects, situations, and organizational contexts. We both enjoy working with many different kinds of people, and sustainability—which requires a great deal of collaboration, across boundaries of all kinds—has been an ideal topic from that perspective alone.

This is a good place for you to stop for a moment and reflect on your own situation. Think about the specific change project that you're facing—or if you don't have a specific project right now, the general types of change you would like to make happen in this world. Let's begin to adjust this conversation and apply it to your needs.

From our own experience we've learned a few things about change that we think are important. Some might seem more helpful to you than others, at least at this particular moment. If so, focus on those for now, and come back to the others when you have a new challenge that requires a different approach. We believe everything included in our book can be useful to you, but you are the one who decides what elements are most relevant to you, right now.

We have tried to keep this little book brief and crisp. But as in a good and helpful conversation in a café, you need to know how much time you have to spend with us now, and what your questions are. If you, in addition, can be on the lookout for something that is a little surprising or unexpected, then you are in the right state of mind for reading this book.

Let us give you a short overview about some of the questions we consider important, or even crucial, for a successful change process:

What basic beliefs do we have, as change agents, about the process of changing systems? How do these assumptions and beliefs structure our perception of reality, the underlying logic we use to explain our specific approach, the choices we make about processes and tools we select, and last but not least, our relationship with the system we work with? How flexible are we, in terms of being able to challenge our own assumptions? Remember what Abraham Maslow, one of the founders of humanistic psychology, once said: "I suppose it is tempting, if the only tool you have is a hammer, to treat everything as if it were a nail."

What method (or methods) do we have for supporting a system to move from state A to state B? We, for example, often use the *VISIS Method*. VISIS stands for Vision, Indicators, Systems, Innovation, Strategy. We start with a clear vision of where we want to be when the change process is successfully complete. Then we look at reality, using formal, measurable indicators about the status quo of the system (when they are available), as well as informal indicators (such as how people report their feelings or general impressions). We want to understand what parts of the system are trending toward or away from the intended state. Next we look for systemic interconnections between different indicators, to understand how things work and to find the spots where we can make the most successful interventions. At those spots we take quite some time to help people come up with real innovations, specific changes that have the power to shift the system. These innovations then need an equally systemic strategy, to go from good ideas to realistic steps forward. That's our (usual) method. Do you have a method (or methods) that you usually apply? What is it?

What is our relationship to the system? Research into different change initiatives indicates that one of the biggest success factors is the quality of relationship between the change agent and the system itself—the people, the stakeholders, and the patterns that need to be changed. Most of the time, change means moving to an unknown and unfamiliar new state. The current state, even when it's not satisfying, at least provides a sense of secure familiarity. People have managed to adjust to it and have found their routines to cope with difficulties or things they don't like. To be willing to leave this behind and move into the unknown, you need to trust your guide. So as a change agent you need to know how to actively create the conditions for that trust. (Please don't misunderstand us here. We don't mean that you need to create a cozy feeling of liking and being liked. We will dig more into this question of trust later.)

How can we increase the capacity of the system not just to change but to improve performance? The new state will not necessarily be better just because it's not old. In fact, the more fundamental differences you can see between the old and the new, the more you need to make sure that the individuals and groups you're working with are striving to be at their best. That means you, as a change agent, need to know how to coach people toward their best performance. This certainly applies to the members of your change team, but it might also apply to the stakeholders you are working with (who need to learn new skills quickly) or even to the CEO of the organization (who needs to energize and inspire the whole company). Successful change and outstanding performance go hand in hand.

How we view change, how we structure our work, how we personally relate to the system, and how we support high performance: these are the fundamental questions that we encourage people to consider, whenever they want to make positive change happen.

Of course, these are not the only questions to ask. Consider your own experience as either a change leader or a participant in a change process. What other questions are important? What else do you believe about change?

3

What We Believe about Change

KEY MESSAGES

- Why do some people see the glass as half full, and others see it as half empty? And how should we, as change agents, relate to that?
- We introduce Appreciative Inquiry and distinguish it from "Repair Mode."
- Here are seven different ways to work with change—and some tips on how to decide which one is the most appropriate in each situation.

Let's order a nice cappuccino or tea in our virtual café, before we begin to explore how our own belief systems, mental models, and frames of reference determine the success of what we want to achieve as change agents.

As your companions, we suggest that we slow down and relax at this stage. Take a good sip of your cappuccino or tea, and taste it, with the fullness of your senses and awareness.

There is an old Chinese saying: "The fish is the last to discover water." In a similar way we are usually the last to discover our own basic belief systems. These belief systems work like a computer program in the background of our awareness. They structure what we perceive, how we filter and weigh information, and how we connect the dots to come up with our conclusions.

Our belief systems structure not just how we see the world but also how we act and react. Behind every individual human decision and action that you observe, in the organization or community that you are aiming to change, there is a belief system, humming away automatically. If we fail to understand that reality, and deal with it actively, then our chances of success are greatly reduced.

Here's a simple example. Some people tend to see a glass that is filled to 50 percent of its capacity as half empty, while others see the same glass as half full. The amount of water in the glass, in both cases, is the same. It's only the beliefs people have about it that are different.

Now imagine how these two very different attitudes might affect the atmosphere in a room, if you are trying to win people over and get them to follow you on a change journey. Obviously an approach that works with glass-half-full people will not automatically work with glass-half-empty folks. Oftentimes you have to adjust your approach to the prevalent belief system.

But sometimes you can succeed in getting everyone to see the *potential* in the glass, regardless of whether they see it as half full or half empty. At the heart of Martin Luther King Jr.'s most famous speech is the phrase, "I have a dream." For example, he said: "I have a dream that one day on the red hills of Georgia, the sons of former slaves and the sons of former slave owners will be able to sit down together at the table of brotherhood." Notice that he did not say, *you* have a dream. He knew that not everyone saw the world as he saw it. Nor was he denying what other people saw and experienced every day: the reality of inequality. But by painting a clear picture of a desirable future state—a *vision*—he was helping to direct people's attention to the potential of the glass's eventually being filled.

To return to our analogy of the fish, there are basically two ways to help a fish to discover water. One is to offer an outside perspective and just tell the fish, "You are swimming in something called water." In other words, we give the fish some feedback about reality and hope that the message gets through.

The other way is to help the fish become more sensitive to its environment and to develop a sharpened awareness of reality. Then the fish will see the water for itself.

As you might guess, we use (and teach others to use) both methods in our work. Self-generated discovery usually has more impact on deeply held attitudes and habitual behaviors. But presenting people with some new facts can also nudge them on a path of new thinking and eventual change—if they trust the source of those facts. Trust will be a recurring theme throughout our conversation about change.

Thinking about How We Think about Change

In addition to the difference between discovering reality through your own experience and being presented with it by someone else, and between seeing a glass as half full and seeing that same glass as half empty, there are many other ways to think about change. To talk about ways of thinking, we often use such phrases as "mental model" or "frame of reference," meaning the concepts, beliefs, and experiences that people habitually refer to to give their perception of reality some clear definitions and boundaries. Following are examples of different frames of reference about change. We suggest that you reflect on which one of these is closer to you. But we also encourage you to ask some close friends or colleagues about how *they* see *you* with regard to these mental models.

Axel: Some time ago my daughter told me that she needed to learn the multiplication tables from 1 to 10. The way she asked me to help revealed that she somehow was coming from the belief, "I am not good in math, and I will never like math." Although she is a very bright and intelligent girl, math is already a kind of nightmare for her. So when I asked her, for example, "How much is 7 × 6?" I could see her muscles tensing up. She was holding her breath and waiting for an insight. But it didn't come.

We were actually facing two different issues. The first one was about learning the correct answer. The second was about helping her strengthen her own belief in her abilities. For me the second issue had the higher priority.

So I prepared a spreadsheet with the numbers 1 to 10 on two axes. We started with those she already knew. 1 × 1 = 1, 2 × 1 = 2, . . . 10 × 9 = 90, 10 × 10 = 100. To her surprise, she discovered that she knew about 80 of the 100 possible answers. She also found out that the 20 answers she did not yet know were actually only 10 different questions, because, for example, 7 × 8 = 56 but also 8 × 7 = 56.

We wrote the combinations that she didn't know on little cards for her to memorize. It only took her about an hour to learn these last multiplication facts. Two days later she came back from school, very proud, and told me that her math teacher had tested her and she had zero mistakes.

Now you might say, "That was really an easy change process, not comparable to the real world challenges I face." And you'd probably be right. But there are some hidden issues behind this story.

Let's consider a challenge we have faced in the "real world." We were asked to help a national government agency that wanted to foster a green(er) economy. (This is a long-term project that is still under way, so we don't know whether the outcome will be successful.) The country in question is in Africa, and it is no longer considered a developing country, but like many similar countries, its economy is very far away from being green.

Put yourself into our shoes: how would you approach an issue like this?

Some of the basic approaches we considered at the beginning of the process were the following:

- Can we create a good stakeholder dialogue, bringing together decision makers from the public and private sector, as well as from civil society, to seek consensus on common goals? (These different groups were not yet talking to each other very much. In dialogue processes such as these, people often find that they are facing similar challenges that can be dealt with more easily if forces are united.)
- Can we do some analysis and mirror back to them that there are already a lot of very successful initiatives with a focus on green economy, in agencies like theirs, taking place within governments not so different from theirs, all around the world?
- Can we focus their attention on some of the things they've already accomplished, show them that they're already moving toward green in ways they hadn't recognized and help them expand on that?

At this writing, many other elements are happening in connection with this long-term project. But maybe you can already see the basic similarity in our thinking between the case of Axel's daughter's learning math and approaching a national issue of economic policy. In both cases we prefer to *build on strengths* instead of dwelling on weakness and *focus on solutions* rather than problems.

Why don't you take another sip of your tea or cappuccino and reflect on your own tendencies when thinking about changing something. Do you usually get drawn into the problematic side, believing that "problems need to be solved" to make things better? Or do you prefer to focus on the existing assets and capacities, and try to make them spread and grow?

There is a comprehensive change philosophy called Appreciative Inquiry that was developed with an emphasis on this second way of initiating change. According to leadership and organizational development expert Gervase R. Bushe, "Appreciative Inquiry (AI) is a method for studying and changing social systems (groups, organizations, communities) that advocates collective inquiry into the best of what is in order to imagine what could be, followed by collective design of a desired future state that is compelling and thus, does not require the use of incentives, coercion or persuasion for planned change to occur."[2]

This frame of reference is so different from a more mechanistic view about change, which seeks to identify what isn't working and then tries to fix it. That approach is certainly not wrong! When building or maintaining a car, for example, it's very important that we find the things that aren't working properly and change them. Even if you drive a $100,000 Mercedes-Benz, one nonfunctioning little part, worth just a few dollars, might stop you in your tracks or even cause an accident. A mechanistic approach, one that involves preventive maintenance, is definitely the right way to keep your car running well.

This "just fix it" style of addressing identified problems shows up in many contexts. Consider a stream that is being polluted by the toxic runoff from a factory. One way to address that issue is to show up at the gates of the factory with signs and banners, demanding that the factory owners stop the flow or runoff, or even close their operation. And there are certainly times when such an approach is called for. But the factory owners might "fix" the problem by shifting that toxic emission to another location.

However, what if it were possible to engage the owners in a dialogue that used the identified problem of toxic runoff as an opportunity to look at the whole production process, and that led to a new, safer, and better way of doing manufacturing? Achieving this potentially more desirable outcome requires a different way of framing our approach to making change.

Here is an example: Some time ago we were working with a manager who was running a team in a change process. The manager complained about two people on his team who were always grumbling and critically questioning his approach. The manager was so stressed about those people that he wanted to ask the top management to replace them. If you have a mechanistic view, this makes a lot of sense. You replace the "parts" that are not functioning well. Instead, we encouraged him to listen more deeply to what these two people were talking about when they voiced their criticisms. This turned out to be an eye-opener for this manager; he discovered that their critiques were shared by about 70 percent of the whole organization. Suddenly these two "grumblers" became very useful: if he was not able to win them over, how could he succeed in the large change process? The manager learned to see the benefits of listening to those two people and used them actively to refine his plan. Instead of replacing them, he gave their point of view an important status—and the whole team benefitted.

But in many Western and industrialized cultures, our minds have generalized the "just fix it" strategy to the point where we simply miss opportunities like these. Many of us believe that change is something we do because something is flawed or not working well. Change is linked to problems and deficits. We are not good, and we must change to become better. The glass is half empty.

Please take a little moment for self-reflection, and think about a couple of change processes you were involved with in the past. This could be on an individual level, such as trying to help your child to improve at school, or supporting one of your staff members to grow into a new area of responsibility, or trying to get yourself into better physical shape. But it also could be in an organizational change process, or in an activist campaign addressing an issue you care about.

- When you think about that experience, what jumps into your mind first? The problematic side of the issue or the well-functioning side?
- To make things better, do you tend to believe that you need to work on the deficits or focus on the existing strengths?

- If you were to give the people participating in the change process (including yourself) spontaneous feedback, how much would you tend to focus on the positive and how much on the negative?
- If you asked for feedback to help you grow, would you respond better to hearing about what you're doing well or about what you're not doing well yet?
- Do you believe it is possible for people to grow and improve if you focus on their strengths?

Imagine how differently a change process would be organized if the change agents viewed the given situation through their "repairing problems" spectacles compared to their "Appreciative Inquiry" spectacles. How would the people involved in the change process feel, react, and contribute?

Is the second approach "better" than the first one?

Alan: Once I was involved in an interesting experiment. At the annual meeting of an international association, we ran two parallel one-day workshops. Both workshops were focused on exactly the same questions: How could the organization sustain itself in the future? What strategy should the association adopt? But each workshop group used very different methods.

One group used the VISIS Method, working especially with indicators and systems analysis. They listed problem areas as well as strengths but focused more on the problems. They analyzed the system dynamics that contributed to those problems, identified where change could be made, and brainstormed new approaches and initiatives to apply to those spots in the system. Then they developed strategies for implementing those changes.

The other group followed an Appreciative Inquiry approach. They did an inventory of "the best of what is" about the association. They dreamed up their common aspirations and visions. They designed avenues for achieving that dream, focusing on possibilities rather than challenges. And they thought together about how to make those avenues reality.

At the end of the day, the two groups compared their results. They had arrived at exactly the same conclusions! Both groups seemed to be entirely satisfied with the process they had gone through (they had self-selected which process they wanted to participate in).

But I noticed—and I was the one who led the VISIS Method workshop—that the people from the Appreciative Inquiry group were smiling more.

As the story above illustrates, we don't believe that there is a single right way to pursue change. There are many possible ways. The trick is in knowing which way is most likely to produce good results—including the result of how people react to the process.

The VISIS Method and Appreciative Inquiry are two commonly used approaches to change, especially in the sustainability circles in which we move. They can arguably achieve similar outcomes, though they might appeal to different kinds of groups. And there are many, many other specific approaches to change, of course. In fact, it's a bit of a jungle out there.

To navigate that jungle we use a model that we have adopted from management consultant Léon de Caluwé, to differentiate between seven *general* ways to approach change. Most of the *specific* methods that we know about fall into one of these seven categories (or some combination of them):

These seven ways lead to very different ways of looking at any given situation. They lead to different planning actions and different ways of relating among the people involved.

Let's elaborate a bit on the differences.

Following a Plan: Anyone familiar with project management knows this approach very well. Generally you start by describing the intended future outcome, as precisely as possible. In an orga-

Seven Ways to Approach Change	
Following a plan	Map a path from *now* to *later*: Think, do, check, and deal with problems/resistance
Enforcing the future	Install a system of carrots and sticks, incentives and punishment
Negotiating an outcome	Play the politics: "What's in it for you?" and "What's in it for me?"
Creating the future together	Participatory development: meet, talk, listen, and decide to act, together
Enabling spontaneous growth	See what is sprouting or flourishing already, and support it
Being visionary	Plant seeds for change with dreams, images, and emotions
Facilitating generative dialogue	Use deep conversation to let go of old patterns and enable new understanding to emerge

nizational context this description often comes directly from the leadership. The organization wants to become more efficient, grow by X percent, implement a new software package, build a new warehouse, install a recycling system, publish a sustainability report. . . . The task of achieving that outcome is given to a responsible project manager or to a team, and they sit together to draft a clear project plan. They identify resources, assign responsibilities, set milestones, and create a mechanism for monitoring progress.

Enforcing the Future: This model had its origin in the psychological school of behaviorism. The focus is on behavior, and the core belief is that if we support desired behavior with incentives and praise, and block unwanted behavior through punishment, people will tend to behave in the desired way. All incentive mechanisms in companies (giving bonuses, for instance) follow this way of thinking, but so do market mechanisms, such as emissions trading:

A company is rewarded monetarily when it reduces its pollution because it can sell excess emission rights and is punished when its pollution rises.

Negotiating an Outcome: When negotiating, people come together to search for a common future that will not disrupt their individual interests and trajectories too much. This model for approaching change is often used by representatives who stand for the interests of a large group. Most political decision-making processes seem to follow this model. Consider the big United Nations climate change conferences: Political representatives from different countries are bound by their national interests but try to find a way forward together on the international level. They trade concessions and search for clever language and solutions that can satisfy all the actors. If the negotiation process is successful, a stable compromise and agreement is found. This process is often very slow, and the concluding agreement doesn't automatically lead to change, as implementation and even monitoring are often subject to continuous negotiating and the weighing of individual interests against common objectives.

Creating the Future Together: This approach to change is vision-, consensus-, and dialogue-based. It is used when there is a need for real buy in (full acceptance and committed participation) by different stakeholders. Sometimes, for the change to be both optimal and sustainable, everyone affected needs to be involved and consulted throughout the whole process. Say you want a sales department to refocus their strategy on new market opportunities. Certainly this could just be planned or enforced, top down, by the head of the sales department. But if you engage the whole department, and gain their enthusiastic participation, you can tap into much more of their experience and creativity. Or suppose you're trying to make labor market rules more humane in a specific region. If you don't get all the stakeholders "bought in," from the public, private, and civil sectors, the whole initiative will likely be in danger.

Enabling Spontaneous Growth: Sometimes the system we need to change is so large and complex that it becomes impossible to foresee the results of any initiative. In such cases it might be helpful to observe the system closely and try to set change in motion, but give the system freedom to explore its own options. You keep observing, and when positive changes begin to occur (more or less spontaneously), you support and reinforce their growth. This approach assumes that the intelligence of the system as a whole is greater than the intelligence of the individuals who are trying to change it. Imagine, for example, that you are the change agent trying to drive a municipality toward greater sustainability. Trying to plan, enforce, negotiate, or even cocreate change in a wide variety of local industries, schools, neighborhoods, and associations would be far too complicated, especially for your small budget. So you observe where sustainability-related change is already occurring and then announce a little competition, to give the best initiatives a prize and some recognition. This accelerates a process already in motion and stimulates others to come up with new ideas that you, as a change agent, might never have identified.

Being Visionary: This approach focuses more on *being* rather than *doing*. Nelson Mandela did not do very much during his twenty-seven years in prison on Robben Island. But by being who he was, holding a clear vision, and expressing it through his behavior as well as his words, he became a monumentally successful agent of change in South Africa. As the French writer Antoine de Saint-Exupery once wrote, "If you want to build a ship, don't drum up people together to collect wood, and don't assign them tasks and work, but rather teach them to long for the endless immensity of the sea."

But we don't need to be Martin Luther King Jr. or Mahatma Gandhi to use this approach. Being visionary is about winning over the hearts of people with a compelling picture of a better future, so they naturally want to contribute with changes in attitude and behavior. To use this approach we need to be clear about our own vision: Do we have one? Do we really express it fully? It doesn't have to be about changing a whole country; it could be about changing your company, or your department. The visionary approach helps people continue moving in the desired direction, even when times get tough and the inevitable doubts arise.

Facilitating Generative Dialogue: In many deeper change processes, and especially those that require a change in culture, you come to a point where all the reasons and answers people come up with don't seem to touch the source of the issue. In fact, staying at the level of reasons and answers appears to be part of the problem itself. This is an indicator that you are stuck in a paradigm—a deeply held set of beliefs about how reality works—and you won't be able to find your way out of it by using the logic and reasoning of that paradigm. Albert Einstein once said that no problem can be solved by the same kind of thinking that created it. In order to create a paradigm shift, a move to an entirely new way of thinking, you need a specific approach for thinking *together*. Generative dialogue is a way of engaging with others in which you (a) create the ability to observe and reflect on the *way* you are thinking, *while* you think together; (b) develop the flexibility to *let go* of patterns of thinking; and (c) create a sense of *openness* and *attentiveness* that allows new patterns to emerge and be recognized.

Obviously you can blend different approaches to change, and if we were to keep digging, we might come up with some additional approaches, too. But sometimes it's good to know when to stop cataloging and start reflecting.

As your companions, we suggest that you take some time and consider this question: Which of these seven approaches do you use most often, or feel yourself gravitating toward most strongly?

Each method has its strengths in specific situations, but they all have limitations in other circumstances. They are all tools, and it's good to have a varied toolbox. Remember Maslow again: if the only tool you feel comfortable using is a hammer, you might start trying to squeeze every reality you see into the shape of a nail, so that you can hit it with your favorite tool!

A second question: Can you think of a situation where you could try out an approach that you don't often use? Then you will become a little more familiar with a different mental model of how change happens.

After all, changing your own belief systems about change might take some practice.

4

Preparing Yourself to Begin

KEY MESSAGES

- The first step in any change process is the most important one. It sets the pattern for what comes later.
- Before you start, map your belief system, and the belief system of your stakeholders, with regard to change.
- In your plan, be sure to include some time for "changing the way we think about change."

Here we are again, in our little café. In our previous conversations we were reflecting on how to approach the topic of change, in a general way. Now we are going to get a bit more practical. How should you start? What should you actually do? To think about this question, it's still important to keep the contents of the last chapter in mind, since your approach to change influences every decision and action you will take later on.

We assume that you have a change process in mind, that you need to plan or that you are involved in. So to begin, reflect on the following three questions:

- What is your basic belief system about inducing change?
- What is the basic belief system of the stakeholders; that is, the people involved in the change process?
- Which approach should you focus on, to achieve the results you intend to create?

With these questions and your specific change challenge in mind, use the Personal Approach to Change template to clarify your own frame of reference for addressing that challenge. On one axis you find the seven ways of looking at change. On the other axis you see the general belief regarding whether the focus is on the existing positives (what's working well) or the problematic negatives (what needs to be fixed). In this table be honest with yourself, and indicate where you tend to spend most of your thinking time, when it comes to planning for change.

You might notice that your own belief system is more like a cocktail, made out of different ingredients. In that case, rather than just checking boxes as you think about them, you could estimate the percentages of each ingredient in your personal cocktail, so that

Personal Approach to Change

	Fixing problems	Building on the positive
Following a plan		
Enforcing the future		
Negotiating an outcome		
Creating the future together		
Enabling spontaneous growth		
Being visionary		
Facilitating generative dialogue		

they add up to 100 percent. (For example, maybe you tend toward 80 percent of one belief system, but that's mixed with 20 percent of another.) Please be creative, and use the template in any way that helps you gain more clarity about your approach to this whole topic.

Here's an example.

Axel: I am currently consulting with a legal practice. They have grown from a start-up two-person company to an office of forty-five professionals. In the past the organization has focused mainly on professionalizing the core competencies of its lawyers and didn't take too much time looking at how to create the right structure, the internal processes, a leadership philosophy, and so on. When they were smaller, everyone somehow knew who was doing what and what needed to be done next. But at the current size this is no longer possible. As a result, they all feel that they are not as efficient as they could be, and the morale has decreased because of this. More people are leaving the organization than in the past.

Using our template, I chose first to put about 60 percent of this group's energy into the box called "Creating the future together." It was important to win people over and motivate them to shape their working environment. We were definitely focused on the positive and were framing the process as moving from "good" to "great."

Then we used about 20 percent of our energy to focus on "Being visionary." Still keeping the positive in mind, we even hired an illustrator to attend the first couple of meetings and convert their new ideas for a positive future into an image.

The moment we had a clear vision, we invested about 20 percent into "Following a plan." This plan was decided by the whole law practice and also managed by them all. We identified those aspects that were not working well, but we also looked for already well-functioning examples. The whole process was organized as a combined top-down and bottom-up approach. The voice of the owner definitely had a stronger weight, but it was never really overruling the others. The owner knew that only together could they reach their goal.

Once you have reflected on how you currently think about change, and you have a sense of which approach would work best in your current situation, there are a few more questions you will need to answer for yourself:

- How does your belief system match up with the belief system (or expectations) of the stakeholders? (What does their mixture of ingredients look like?)
- Do you need to work with the stakeholders first and help them understand the approach you plan to take?
- Do you need to change yourself and your own belief system first, before you begin work on changing others?

If the answer to either of these last two questions is yes, then be sure to include those steps in your planning!

One of the most important steps in doing personal development is becoming conscious about *what* to develop. Mapping your own preferences toward "Personal Approach to Change," as described in the table earlier, gives you the opportunity to look into a mirror. Maybe you can already see here that the specific approach you described is unique for this current change process, but maybe you already notice that this is your *habitual* way of dealing with issues like this. Often it is hard to differentiate these two. In those cases, it is a good idea to start talking to other people, friends and colleagues. How would they approach a change issue like this one? And do they notice that you are usually doing things "your" way? "Your" way is always an indicator of patterns and habits.

If this is true, you might want to reflect a bit more on how to approach your own development. Maybe you'll need to learn about other frameworks for change (such as Appreciative Inquiry) in more detail. Maybe you'll need a bit of training, or coaching, to help you shift your perspective and your habitual way of looking at change. And maybe you'll need to include a step whereby you introduce these ideas to the stakeholders as well and listen carefully to their reactions.

Remember, this first step is very likely to be the most important one. Being conscious about your approach, and matching that approach to the people you are working with, will help set a tone of clarity and confidence from the outset and boost your chances for success.

Of course, not all change processes take place in an organizational context. In the next chapter we will look at a model for change that will help you move your ideas forward, whether you are working in an organization or with looser groups and networks of people.

5

Change Is Several Processes at Once

KEY MESSAGES

- As change agent, you are usually managing many different process challenges, in addition to the contents of your intended change.
- You are also managing yourself as a change agent.
- In that complexity there are ways of improving your chances of success, by considering three key dimensions: stakeholders, planning, and finding the right level for intervention.

One of the biggest challenges in managing a change process is the complexity. As a change agent, you are not just doing one thing: pushing an idea forward and getting some new content into the system. You are actually handling several *processes* at the same time.

First, there is the *communication* challenge: you have to get the idea (or usually multiple ideas) across to a variety of different target groups, in a compelling, persuasive way.

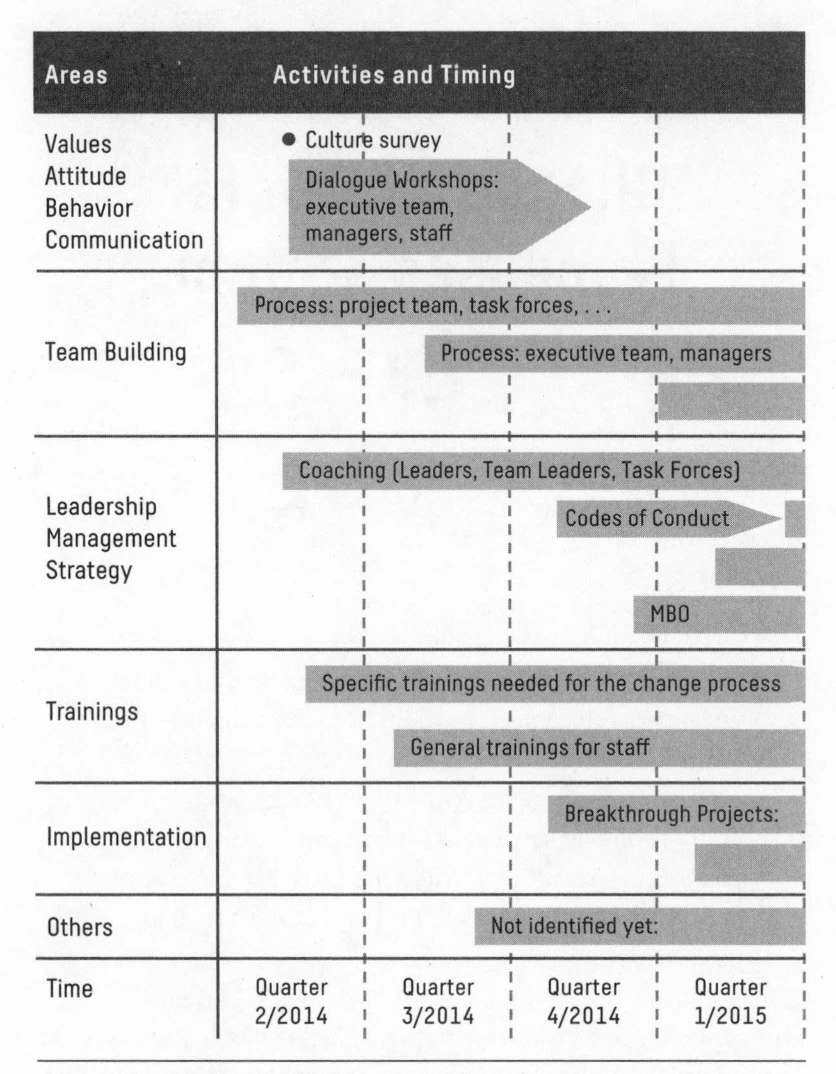

Cultural Change Process. Sample of a chart for mapping a complex organizational change process. Courtesy of Axel Klimek.

Second, there is the *implementation* challenge: you have to give people a clear pathway showing how this change is going to work in relation to the current system.

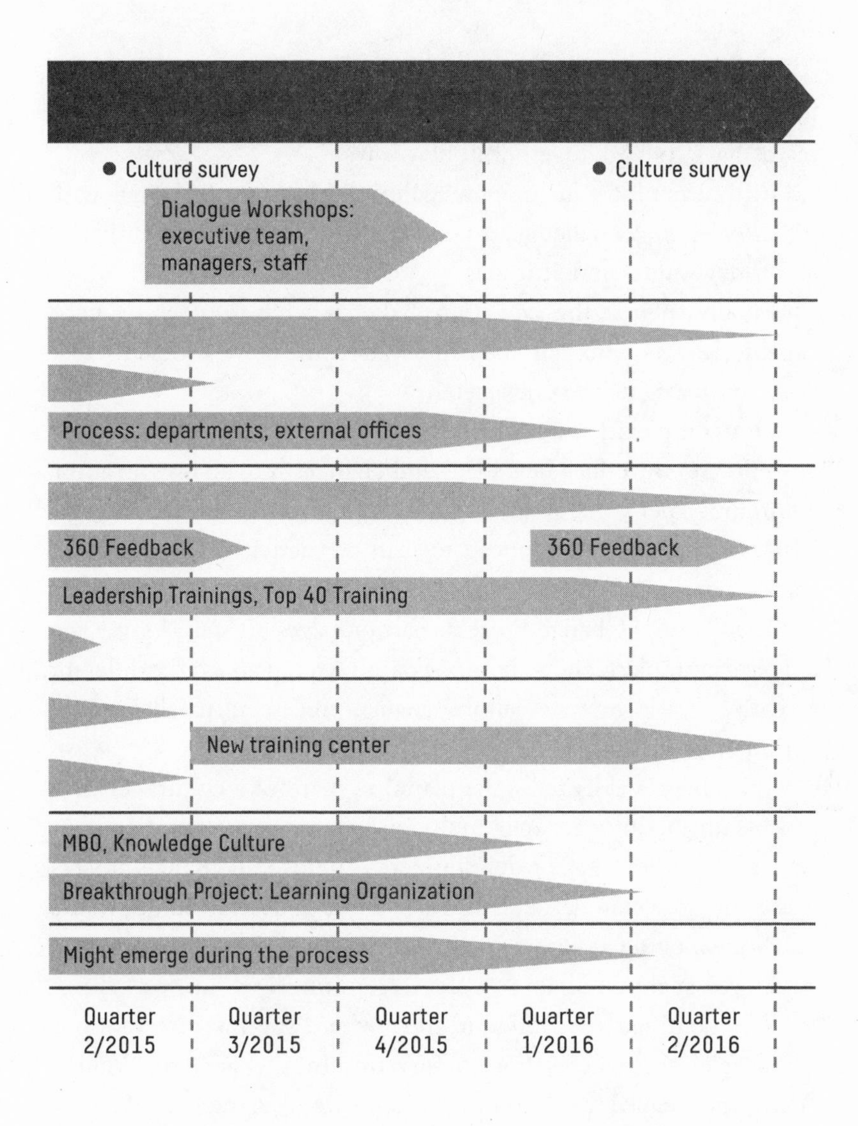

Third, you usually face a *training* challenge, because somebody—or a team of somebodies—needs to understand this change thoroughly so they can implement it.

Fourth, that means you have a *team-building* challenge, since the new way may involve forming new groups of people. Even existing teams have to work together in new ways.

Fifth, even introducing new technical ideas usually brings with it a *culture change* challenge, as people need to behave differently and adopt new values and attitudes.

Finally, there is the *leadership* challenge: change processes need capable leaders who can keep motivation up, solve problems that arise, and navigate the unexpected.

This complexity is a normal, but usually underestimated, part of any effort to introduce new ideas and changes into an organization, community, network, or any other social grouping. And this means that as change agents we need to plan our activities along parallel tracks that are interlinked.

The Cultural Change Process chart on pages 30 and 31 gives you a typical example. It shows how one of us (Axel) planned the different activities in a large-scale cultural change process, in parallel, over a period of two years.

This chart is essentially the planning map for a cultural change process in an international multigovernmental organization that was in its early stages of development. The overall change process was even more complex, and this is just one of several similar charts that were looking at necessary changes at the levels of processes and structures, such as human resources, recruiting, and finance.

We identified the different areas where specific interventions could take the whole process forward. In the area of "Values, Attitudes, Behavior, Communication," we designed a series of Dialogue Workshops that helped everyone involved to find a shared understanding about the organization they were trying to develop. For the success of the whole change process, different teams were very important: the top management team, the project team (the task force working on the change process), and the teams working in each department. To boost their energy and get buy-in, we conducted separate team-building events. On the level of leadership and management, we conducted "360 feedback" processes, in which managers receive open feedback about their leadership style not only from the people they are leading but also from peers and their own supervisors. In addition, we helped them develop a management by objectives (MBO) approach and describe a code of conduct for the organization.

To help all staff become up to date on their own skill sets, we gave them special training as well. Very important projects were addressed through a specific approach called "break through," which is designed to create fast, tangible results, accompanied by a learning and reflecting mind-set. As the chart shows, it was important not to do these activities one by one, but rather to start many things at the same time and continuously weave things together.

Actually, the complexity does not end there. Those are some of the key challenges we face as change agents. But there is also another, central challenge that we can only really notice if we step back and look at the whole picture . . . including us.

As change agents we also face the challenge of *managing ourselves*—building our own capacity to handle this complexity; making the right strategic choices; keeping the parallel processes on track, with the help of good tools and methods. And much more besides.

Yes, it's a lot to manage and keep in mind! But it is entirely possible. Despite the data we cited earlier about how many change processes do not end in success, quite a significant percentage of change projects do succeed. We would like to help you, and your change projects, end up in that group of successful initiatives.

In the following chapters, we would like to break the challenge into three key dimensions and share with you some tools and approaches that we have found to be very helpful. The three dimensions are:

1. **How to engage with the stakeholders.** The larger the change process, the more people and stakeholders are involved. Without knowing how and when to involve whom, and how to

deal with the dynamics that come up among the different stakeholders or people involved, your chances for success become more of a gamble. To help you, we will focus on two simple reflection tools: the "Amoeba of Change," developed by one of us (Alan); and the "Stakeholder Dialogue Spiral," developed by Petra Kuenkel of the Collective Leadership Institute.

2. **Designing a good plan.** There are many different tools and methods to help you come up with a compelling plan to bring the change you seek forward. As an example of one way, we will run you though the structure of our Pyramid Process, built on a method called VISIS: Vision, Indicators, Systems, Innovation, and Strategy.
3. **Intervening at the right level.** Underneath the obvious reality of a problem you are trying to solve, or an improvement you are trying to make, there are deeper levels: hard-to-see structures, patterns, or processes that give rise to this more visible and obvious reality. Often, if you don't manage to intervene at one of these deeper levels, the intended change will not last. To reflect on this more subtle way of understanding change processes, we will focus on two little models: "Theory U," developed by Otto Scharmer; and the "Strategy-Culture-Leadership Model," developed by Peter Hawkins from Bath Consultancy Group.

These three dimensions—stakeholders, planning, and the different level of change—are very much interlinked, and they influence each other constantly. Also, the tools and methods that we will be focusing on can be used separately, but they can also be mixed and integrated in various ways.

Still, to keep things easy and conversational here in our virtual café, it makes sense to look at these things one by one. We trust that you will be able to see how they fit together and that you will know how to use them in the way that is most helpful to you.

6

Mapping Your Social Environment

KEY MESSAGES

- Different people play different roles in any change process.
- The "Amoeba" model is a tool for mapping those roles, and more effectively planning how to introduce your change process.
- It will also help you figure out your role in the change process, so that you can link up with others who play other roles.

We have some good news and some bad news.

Let's take the bad news first. When you approach a change process, you can be sure that some people are going to be less helpful to you than others. Some will want to slow you down. Some won't care. Some might believe that change is too difficult, or even impossible, and not worth the effort. And of course, some people might actively oppose you. They might work hard—visibly or behind the scenes—to try to make your initiative fail.

It's best not to ignore these uncomfortable facts. But here's the good news, and it has two parts. First, social systems usually balance themselves out to stay stable. So if you find people who are likely to be against your change initiative, then you're also likely to find people who are for it, and who might even have waited for quite some time to have a chance to be involved in such a process. (These people are usually great fun to work with!)

The second part of the good news is that there is a model for thinking about change that can help you prepare for these dynamics, and it also offers some possibilities for dealing with them. This model can also help you plan your approach so you maximize the chances for success, and minimize the amount of time and energy you might otherwise waste by working with the wrong people, in the wrong way.

We call this model "Amoeba."

The Amoeba model was developed by Alan many years ago, and you can read about it in detail in his books *Believing Cassandra* and *The Sustainability Transformation.*[3] The model has a number of parts, including a simulation role-playing game that we use to train change agents. But in this chapter we're just going to give you a short, simple introduction, so that you can start using this model immediately to map your social environment.

Why "Amoeba"? Because the model takes the single-celled creature called amoeba as its metaphor. The model imagines organizations, groups, networks, and whole cultures of people as Amoebas that take in new ideas from their surrounding environment, like food.

Within each Amoeba, there are different people playing different roles, and those roles might change in relation to any new idea or change that comes into their system. They might promote the change, and help it to spread further. They might passively observe the change, only participating in it or adopting it themselves when others around them are doing so (or when they're given no choice). They might actively oppose the idea, or even see it as threatening.

For a change agent, understanding the Amoeba can be very empowering. Once you can visualize these relationships, you can

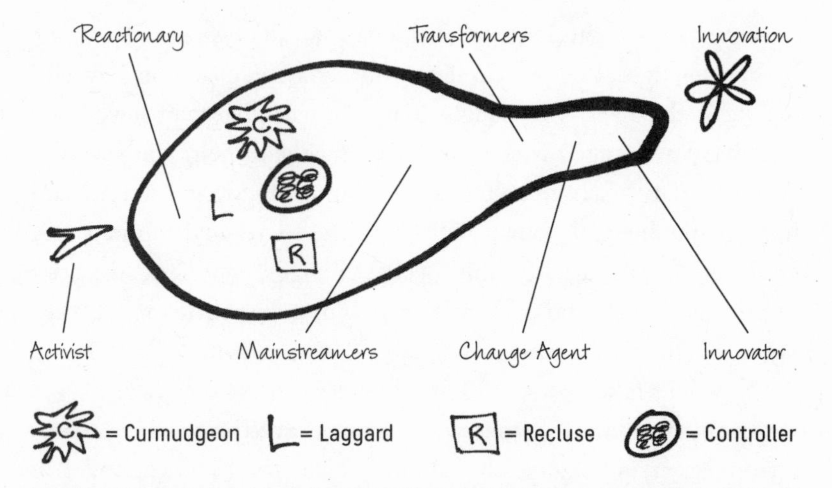

The Amoeba Map. The "Amoeba" model illustrating common roles in the organizational and cultural change process. *Innovators* push the boundaries of the organizational Amoeba to reach a new *Innovation* and bring the new idea inside. *Change Agents* help the Innovators by adapting the Innovation and communicating it to opinion leaders and decision makers, called here the *Transformers* (because they determine whether an Innovation makes it inside, to where it can transform the Amoeba). *Mainstreamers* fill the main body and follow the lead of the Transformers. *Reactionaries* actively resist the new idea, or even push in the opposite direction. *Activists* sit outside, where they critique the status quo or even attack the Reactionaries. *Curmudgeons* are the chronic complainers and pessimists who don't believe in change. *Laggards* just like things the way they are and are late to adopt new ideas. *Recluses* keep themselves out of the action altogether. And finally, the *Controllers* are the ultimate deciders and rule setters (e.g., boards of directors). Often they just watch the process—as long as they judge it benign. But they can step in to stop the spread of Innovations they find to be threatening—or accelerate those they deem very advantageous.

make conscious choices about how to approach them. You are less likely to be surprised, and more likely to be strategic.

Above, you'll see a small Amoeba "map." This is a general map of the roles in the Amoeba model. Once you understand these roles, you can make your own map, noting the specific people or functions or departments or cultural roles that you expect will be playing different roles in *your* Amoeba. Then you can plan your strategy accordingly.

One note of caution: the Amoeba map is never a fixed thing. People change. They might like your change idea one day and dislike it the next. They might start off working against you but later have a change of heart and start trying to help you. Use the Amoeba map wisely, to help you get oriented and to get started, and to help you navigate your way through the social environment. But as always, be prepared for unexpected changes, as well as surprises: people sometimes end up in roles you don't expect. The map, as they say, is not the territory.

Now let's take a look at each role in turn.

Innovators are the ones who first identify the specific idea to introduce, or the change that should occur: the "Innovation." Are you an Innovator? Do you invent, or just get very excited about, new tools, methods, projects, or processes? If so, you might have a harder time promoting them effectively. Innovators typically are very attached to the ideas they champion. They are "in love" with the ideas, and often they are less flexible about how things should be communicated and implemented. To use sales language, they tend to focus on the features and forget to tell people about the benefits. And they often have a difficult time thinking strategically about how to introduce the change they seek and steer it toward success.

Change Agents are just the opposite. They understand new ideas and feel strongly about promoting them. But they also understand people, communication, and strategies for change. If you are reading this book, you probably are (or aspire to be) a Change Agent in many ways. This means you are (or aspire to be) flexible and thoughtful about how you go about introducing change, helping people understand the benefits, and seeing the process of adoption and implementation through to completion.

Note: Some Innovators can be good Change Agents, too, but it requires a great deal of self-knowledge and discipline to move from one role to the other. Change Agents have to learn to see the change through *other* people's eyes, and adjust their communications accordingly.

Transformers are sometimes known as "opinion leaders," "gatekeepers," or "early adopters." They hold a position of respect or authority within the social environment, though they may not be at the top of the power hierarchy. If they endorse and adopt an idea, others will follow their lead. They understand that introducing new ideas and change processes often brings bigger change than anticipated—even transformation. They also value their role in the Amoeba. So they are careful about what they endorse.

Here's the first key strategic tip: Change Agents need to find Transformers and convince them to endorse the change.

Mainstreamers follow the Transformers' lead, as well as each other. They "go with the flow." If you are a Change Agent, talking with a Mainstreamer can be frustrating: they might appear vaguely interested, but they tend not to commit unless they see others around them doing the same.

Reactionaries oppose the change: they "react" strongly against it. There can be different reasons for their opposition: Maybe they truly believe the change is a bad idea that will hurt the organization (or social environment generally). Or maybe they have something to lose, if the change succeeds. They might be principled opponents, with good intentions, but they might also be clever, underhanded foes who will use any means at their disposal to stop your idea in its tracks.

Here's the second strategic tip: Change Agents need to avoid Reactionaries, at least at the beginning of a change process, and make a plan for how they are going to deal with the inevitable opposition and resistance that Reactionaries put up.

Activists are the critics and protesters. To describe the Activists, we sometimes use the word "iconoclast," which means someone who "breaks the icons" or challenges the prevailing beliefs. Activists call attention to the problems and criticize the people whom they see as creating the problems. Typical examples are nongovernmental organizations (NGOs) such as Greenpeace or journalists who point fingers at what's wrong.

A third tip: Activists can help Change Agents by criticizing the Reactionaries and keeping them busy. But Activists can sometimes point their critical fingers at the Change Agents, too, if they think the proposed change doesn't go far enough in addressing the problems they see.

These are the key "active" roles in the Amoeba. Then there are a few special roles as well:

Curmudgeons are committed pessimists and complainers. They don't believe the change will succeed—or even if it succeeds, it will not make anything fundamentally better. The secret to Curmudgeons is that most of them were once Innovators or Change Agents who did not succeed. Sometimes you can "rehabilitate" them and turn them into Change Agents again, but often they will simply drain the energy from a change process.

Laggards are people who simply like things the way they are. They are reluctant to adopt a change or a new idea, because they are comfortable with what they already know. You can think of them as *very* slow Mainstreamers. They will change eventually, but trying to convince them to change is like pushing on a big rock.

Recluses simply hold themselves out of the process. They have other interests and priorities. They might be very wise—think of researchers or spiritual leaders—but they tend not to get involved in the messy business of trying to implement change. If you can get them engaged, they can sometimes be very helpful. But they are also a little unpredictable.

Last but not least are the **Controllers**. They are the DNA of the Amoeba, the people who make the ultimate decisions about what will happen inside it. In an organization they might be the CEO or the board of directors. Controllers can certainly help a change process to accelerate. But they can also stop it dead, if they see risks, costs, or other impacts that they don't like.

One more tip to Change Agents: If the change you are trying to make doesn't originate with the Controllers (which is most often the case), it is usually better to stay out of their sight for a while—at least until you have built up some momentum with the Transformers. Then if the

Controllers need to be informed, let the Transformers introduce the idea to them. Going to the Controllers as a Change Agent, before you're really ready, is often a big risk to the success of your initiative.

That's the Amoeba in a nutshell. So how should you use it?

First, draw yourself a little amoeba outline, like the one shown in The Starting Point.

Then think about the change that you're promoting. Start making a map: write down names, and draw arrows to show which role, in *your* Amoeba, different people (or functions or departments) are likely to play.

As you build your Amoeba Map, ask yourself these questions:

- As someone involved in initiating this change process, are you a Change Agent? Or are you an Innovator?
- If you're an Innovator, do you have the skill and discipline to play the Change Agent role? (Note that the Change Agent role often requires making improvements, compromises, or other adjustments to the original idea or change process, to get the endorsement of the Transformers.)
- Who are the Transformers, and what will convince them to endorse your idea or change program?

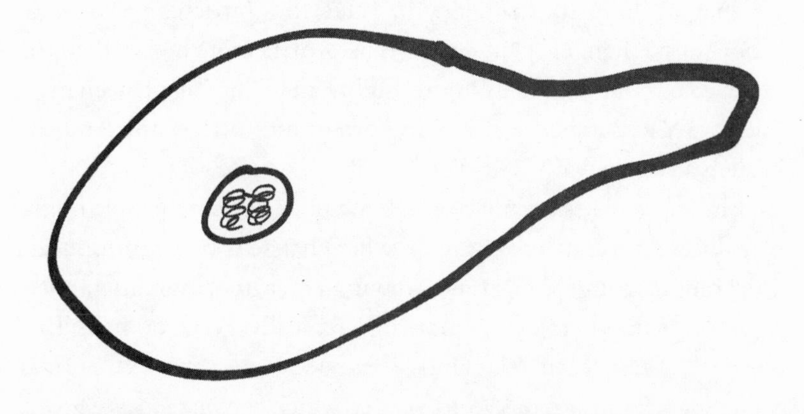

The Starting Point.

- Who are the Reactionaries, and how are they likely to oppose you? What can you do to reduce their impact? (Note that trying to "convert" a Reactionary is often a risky strategy. You might be able to do it, but you might also simply alert the Reactionary to your plans.)

For each role in the Amoeba, make a little plan for how to communicate with them . . . or how to avoid them.

The Amoeba model was first invented in 1990. Since then, it has traveled far and wide and been translated into many languages. While of course there are differences in how different cultures express these things—whether we're talking about national cultures, community cultures, or organizational cultures—the basic roles seem to be universal.

We believe that reflecting on your social environment in this way can strongly improve your chances for success. It has certainly helped us, as well as many other people, to be more strategic in how we plan our change processes.

A Few Important Subtleties about Amoeba

In professional dialogues among change agents, you sometimes see that we have the tendency to value the "pro-change" roles of Innovators, Change Agents, and Transformers more highly than the other roles, and especially more highly than the "against-change" roles of Reactionaries. But it is important not to view any Amoeba in such a simple way.

First, these roles are not fixed. Remember, different people might play different roles, depending on what change is being introduced. They might change their minds about an idea, too, more than once!

Second, these roles are not automatically good or bad. That depends, of course, on what kind of change you're making. (Consider the French *Résistance* in World War II, which can be seen as a group of Reactionaries who were trying to stop Hitler's "innovations.")

Whether a change is good or bad can also sometimes depend not on any objective measure but simply on the perspective of the person or group, and whether they see the change as beneficial to them or to the whole Amoeba.

And third, some of the "against-change" forces in the Amoeba can actually be seen as positive, even when the *change* is clearly positive—especially if they're helping to govern the *speed* of change and preventing the whole Amoeba from being damaged or pulled apart by a too-rapid transformation.

Reflect on these points when you are mapping your Amoeba. In fact, the most powerful strategy is to keep all these roles in mind, with great care and consideration for everyone in each role, as you plan and implement your change efforts. We'll come back to this later.

It's also important to remember that systems evolve in phases. After each phase of change, there follows a phase integration and consolidation—making the change a part of "normal reality." During this integration phase, it is crucial not to initiate the next round of change too quickly, as in this example:

Axel: I was once involved in a big merger process in which two different organizations with very different cultures needed to join together. One group had more of a start-up mentality, and the other was a big, slow, bureaucratic entity. The big and slow part had bought the fast, small one.

About two years after the original merger, we were planning a workshop, using an organizational chart, which had been given to us a few days earlier. When we presented the design of our workshop to the people we needed to invite, we learned that this organizational chart was no longer up to date. In a single week the organization had been restructured—twice; for more than two years the new entity was in a chaotic phase. During this whole time, there was no clear process of "learning from each other." Those responsible kept looking at changing structure and processes but not at how to bring the people—their different mind-sets and different cultures—together. In this organization, even ten years later, you could still see the two different cultures, and the scars from these too-fast change efforts—and they are still struggling to be successful.

In the example above, if perhaps the people who were trying to slow down change had been listened to a bit more, the outcome might have been much better.

What about you? What experience do you have with Amoebas? What roles have you played in different change processes that you've been involved in?

7

Managing Stakeholder Dialogue

KEY MESSAGES

- Everyone involved in a change process has an interest in influencing it.
- Using multistakeholder dialogue increases the chances that people will participate and buy in to the change.
- If you approach it as a learning process, you can build on one success to create other successes.

Let's take a look at the people who are involved in your change process: the stakeholders.

We define a stakeholder as a person or an entity that can affect or be affected by the actions and results of the change process, in all its aspects. Every stakeholder has an interest in influencing the change process, so that it is more likely to meet her or his own needs and wishes.

There is a nice line from a film called *The Legend of Bagger Vance*: "If you can't beat them, lead them." For us as change agents, this means that when we don't have the power to directly create the outcomes we want, we need to figure out how to lead people toward a shared goal.

Petra Kuenkel from the Collective Leadership Institute has developed a process for doing just that: leading multistakeholder initiatives to set and reach goals that all stakeholders are willing to support actively—or at least to not act against.[4] (We summarize some of Petra Kuenkel's ideas here, but more information can be found in her book *Working with Stakeholder Dialogues*, written with Silvine Gerlach and Vera Frieg. We also recommend her book *The Art of Leading Collectively*.)[5]

Here is an example, once again involving a big change process to promote the concept of a green economy in an African country (which we described earlier). In that process we have many different stakeholders to handle, all with different interests. There are the companies that benefit from the old system of a non-green economy. There are the companies that see an advantage for themselves in the new greener approach. And there are companies in the middle. We also have political parties to consider, unions, representatives from the civil society, political activists, communities. . . .

In this process we experienced a challenge that is common in relating to stakeholders. On the one hand, you need to relate to people as individuals, with their own interests, opinions, needs, and wishes. On the other hand, people are often representatives of organizations, and this requires them to align their actions and decisions with the interests of those organizations.

This contributes to making stakeholder processes very sensitive. Sometimes a small mistake can endanger the success of the whole enterprise.

The chief aim of a good stakeholder dialogue is to build a climate of trust among all those involved. This is essential to getting their commitment to be actively involved in the change process (or at least not to intervene against it). A good dialogue creates a kind of "container" in which people move from being individually smart people to working together as a kind of collective intelligence.

Petra Kuenkel describes four different phases in designing a multistakeholder dialogue process, and in each of the four phases there are three aspects to focus on. Graphically, she arranges these in a spiral, to illustrate how the earlier phases are definitive in setting the shape for the later ones.

Now we'll describe the four phases, and you can follow along with Kuenkel's Dialogic Change Model.

Phase 1: Exploring and Engaging

This phase sets the tone for the whole initiative. From the beginning you need to be clear about the content and start winning over important stakeholders who support the initiative. The success of the final result is determined at this first phase.

Step 1: Create resonance

In this very first stage, you start talking to different stakeholders to find out if they think the initial idea makes sense or not, and

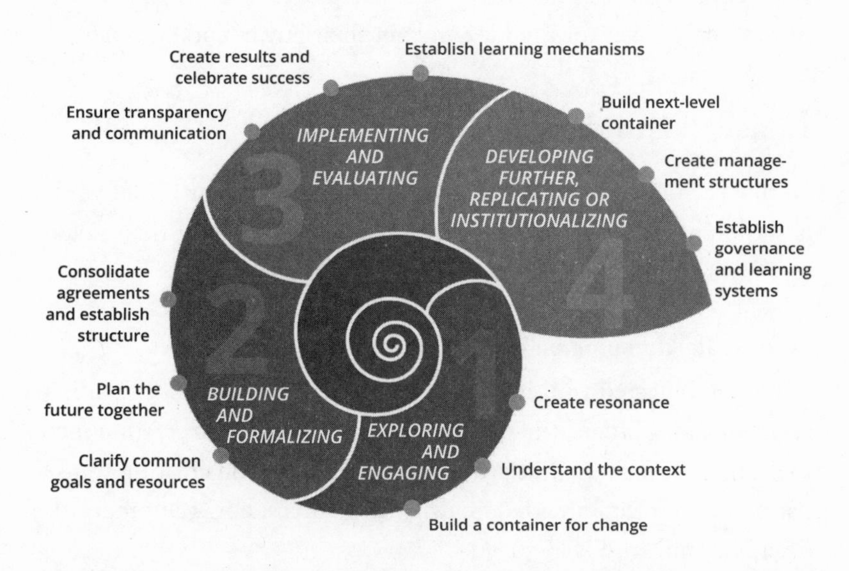

Dialogic Change Model: Implementing Stakeholder Dialogues in a Result-Oriented Way. Petra Kuenkel's Dialogic Change Model for mapping a stakeholder process. © Petra Kuenkel, Collective Leadership Institute

whether they see themselves fitting into the strategies and actions. Doing things "dialogically" means that from the very beginning you are asking questions and listening—not just telling. The "dialogic" approach sets the template for everything that follows.

Step 2: Understand the context

If you find that your initial idea does make sense, then you need to dig deeper to understand the reality: all the possible stakeholders and their interests, potential conflicts, existing initiatives, who supports what, and so on. At this stage it is often important to benchmark whether similar ideas have been initiated in other areas.

Step 3: Build a container for change

With all this contextual knowledge in your back pocket, you start the dialogue process by inviting people to a first round-table discussion (or series of discussions if you have a lot of stakeholders). The more stakeholders you can involve in this process, the better. The aim is to win people over for the idea and get their buy in and ownership.

Phase 2: Building and Formalizing

In this second phase you start going public with your initiative. It is the moment when an idea is tried out in reality and then taken forward to grow, creating momentum and change.

Step 4: Clarify common goals and resources

After you've tested your ideas and your initial process design with a series of stakeholders, the moment has come to "go live." Announce the formal start of the initiative. You should focus on creating transparency about the intended goals and resources, how the process is designed, and who is involved.

Step 5: Plan the future together

From this moment on, your role is more and more about being a facilitator. You should leave space for everybody to participate in shaping how to move forward.

Step 6: Consolidate agreements and establish structures

As in any good project, you now need to create a consolidated plan about who needs to do what, by when. All events and decisions need to be documented and made available to everybody. Often it's also important at this stage to get administrative and logistical support for the process from the stakeholders involved. The more effort people put in, the more likely it is that they will continue.

Phase 3: Implementing and Evaluating (ensuring the smooth running of the stakeholder dialogue)

Every initiative is always evaluated at the conclusion. You want to know if it has created impact or not. So you need to dare to really move out into public view and create the intended change. To keep spirits high over a long period of time, you need to make sure that successes are noticed and celebrated.

Step 7: Ensure transparency and communication

You might notice at this stage of the process that some stakeholders are more active than others. Still, it is important to create a transparent communication system that constantly connects and reconnects everybody into the process.

Step 8: Create results and celebrate success!

The whole endeavor always needs to be framed in such a way that it can create rapid results. A stakeholder dialogue is not intended just as a forum for discussion but as a process for quickly focusing awareness on trying things out in reality and then reflecting on the experience. Success should be celebrated and promoted to keep the momentum going.

Step 9: Establish learning mechanisms

Part of creating a learning process is to continuously reflect on what has been done and on *how* things have been done. So you need to install a feedback culture and a learning atmosphere. As a result of discussing the experience—what went well, where have we not performed well enough—the process might be readjusted from time to time.

Phase 4: Developing Further, Replicating or Institutionalizing

The sign of a successful initiative is when it has become a *habit* of behavior, process, or perception. To achieve this, it needs to grow into a widely accepted piece of organizational knowledge or structure.

Step 10: Build next-level container

Through this reflection process, you might easily come to the point where you realize that the original focus was too narrow, or that there are other issues that also need attention, or that there is the need to scale up. At these little moments you can build on the fact that your original initiative has proven to be effective. You can harvest some further results by developing new containers or involving new stakeholders, and possibly tackling new questions or ideas.

Step 11: Create management structures

Every change process is a kind of incubator. It is only successful if the change that was initiated continues. Thinking about how the change will last, and will be integrated into an existing system, is crucial from the beginning. But this is the stage when that integration should be realized.

Step 12: Establish governance and learning systems

If you have accompanied the change process to this level, there is a lot of learning to be harvested, which can be used for similar initiatives and for the stakeholders themselves, individually and for their organizations. You can also help ensure their future cooperation in other initiatives. This is the moment to critically compare what you *intended* to achieve with what you *did* achieve, and also reflect on how you performed over time.

The last step: Celebrate, and take the learning forward!

* * * * *

We're sure that you can immediately see how this stakeholder dialogue approach can be useful in many situations. There are many varieties to it, as well. For you as a change agent, the key is to bring together skills as a listener, communicator, process leader, and strategist.

But your skills will likely need to be supplemented by the skills (and the energy) of others. When you're leading a complex change process, you usually need to work with a team. Remember the old African proverb: When you need to go fast, go alone. But when you need to go far, go together.

Also, you probably noticed that, at some stages of the stakeholder dialogue process, you can certainly benefit from doing the Amoeba mapping we talked about earlier, and taking those dynamics into consideration. Whenever you think of the whole group of stakeholders, keep the Amoeba in mind. There are always some who are more active than others, more pro or con, and more likely to be seen as opinion leaders (Transformers). There are almost always those who are even against the whole process (Reactionaries)—because from their point of view, there seems to be something to lose.

So how about you? Have you led a stakeholder dialogue before, of some kind? Do you have experience facilitating group conversations among people with diverse interests? If you don't have that experience yet, and you need it, how could you acquire it? Or can you team up with someone who has?

8

A Methodology for Systemic Change

KEY MESSAGES

- The VISIS Method can guide you (or a group) through a process of thinking that leads to new insights about change.
- Even a little systems thinking will go a long way to helping you advance change. VISIS helps get you started.
- Now you can begin to see how different tools and processes fit together. A key skill for change agents is learning how to choose the right tool for the job.

Let's talk about one more method for leading change. As we noted earlier, we often use a method called VISIS, which stands for Vision, Indicators, Systems, Innovation, and Strategy. These are five steps in a process that you can use with any group, or even by yourself as an individual. In a moment we'll show you how you can also include this process in the stakeholder dialogue approach as well. But first let's introduce the method to you, so that you can apply it in your own work.

We use VISIS in two different ways: for training and for planning. Especially when it is used for planning, the whole purpose of the method is to prepare people to take action—and make change happen. We often use VISIS together with a little workshop tool called *Pyramid*, which is part of our *Accelerator* tool kit. Pyramid takes VISIS, combines it with another tool called Compass (which we describe below), and turns it into a group process workshop that leads people through VISIS step-by-step, while maintaining a whole-systems approach to sustainability. Accelerator is our professional

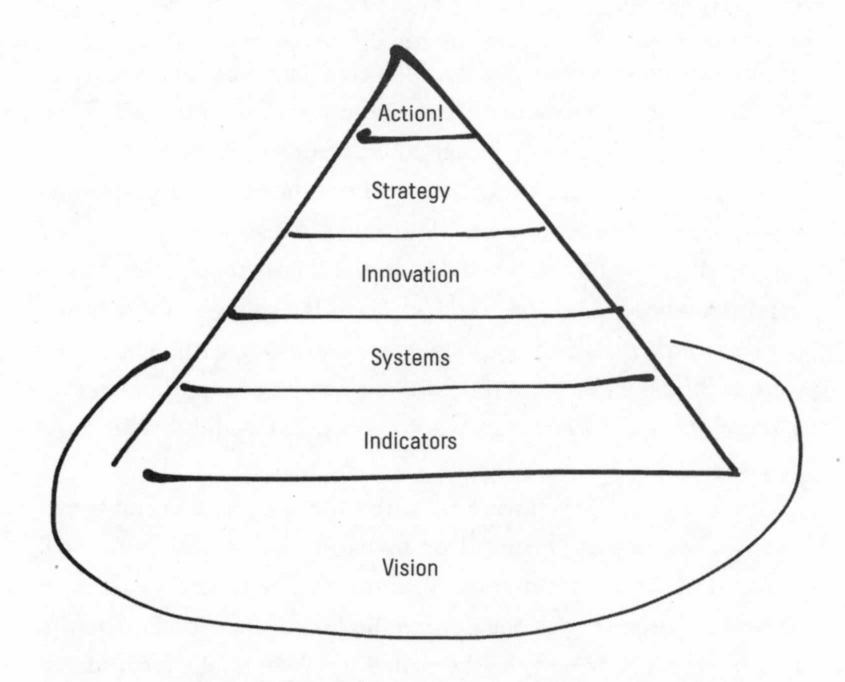

The VISIS Method. The pyramid shape of the VISIS diagram helps to emphasize that the method builds understanding and consensus, step-by-step. Groups discuss the options and prioritize them at every stage, until they ultimately come to agreement on the best course of action. In a VISIS workshop we often build a physical pyramid to reflect the stages in this process. By the time the group reaches the top of the pyramid, there is usually a very strong consensus on how to move forward.

package, which includes manuals, templates, and presentation slides for using our methods in an organizational or educational setting.[6]

But in this part of our conversation, we're going to focus on planning for change. You can use VISIS to plan the implementation of a change idea that you already have or to search for new solutions. The process has a logical flow that produces a lot of information. Then you can build that information into your change process in various ways.

Vision

Before starting a change process, or even a search for a new idea, it's helpful to have a common picture of the hoped-for outcome. Why are we doing this process? What do we want to achieve?

A vision can be anything from a short statement of purpose to a fully developed description of the new situation you intend to create, with lots of details. You decide just how extensive this part of the process needs to be, based on what kind of change you are trying to create and what kind of group you are working with. Sometimes you need to spend a long time articulating a vision; other times it is more effective to just start with a general idea and dive into the process.

For example, are you working with experts or leaders and trying to make a complex technical or managerial change? Then you'll probably need to spend some time making sure the specifics of your vision—or at least the conditions for future success that you are trying to meet—are clearly spelled out. We usually recommend in such cases that you start from a clear set of scientifically sound principles for sustainability, too, and make sure your future vision aligns with those principles.

But what if you're working with a mixed group that doesn't yet share a detailed vision, such as a community that's very divided on the issue you're working on? Or what if you're trying to find your way to a better situation, but you cannot really describe your vision yet? In cases like these you should probably *not* spend too much time trying to create a detailed vision: a general statement of purpose or intention is enough to get going.

Hint: in those latter cases, using this VISIS method will help you *create* a more detailed vision.

Indicators

Once you've identified the vision or purpose of your exercise, you start gathering information. What are the key issues and trends that we are dealing with? What data do we have to look at? Where are things currently heading? Are we moving toward our vision (even if the vision is very general)—or away from it?

Note that you should not limit yourself to trends that can be backed up with data at this stage. "Indicators" are signals of any kind, telling us about what is happening. For example, even if you are not measuring "employee engagement" or "satisfaction" in an organization, you will probably know whether people are generally happy. You will probably know whether their level of engagement is increasing, declining, or staying the same. Even if you personally don't have a clue, you can ask around and gather informal impressions, which are also indicators. (But if you work in an organization and you do not have a clue about whether people are generally happy there, that in itself is a worrying indicator!)

Alternatively, you can look for what we call a "proxy indicator": something that's indirectly related to the thing you want to know. For example, when employees are happy and engaged at work, they take sick leave less often. The trend in absenteeism will probably tell you something about employee engagement, satisfaction, and happiness as well.

To be sure that we are complete in our survey of the issues and trends that might be affecting the situation, we use another little tool called the Sustainability Compass. We take an ordinary compass, with its directions north, east, south, and west. We keep

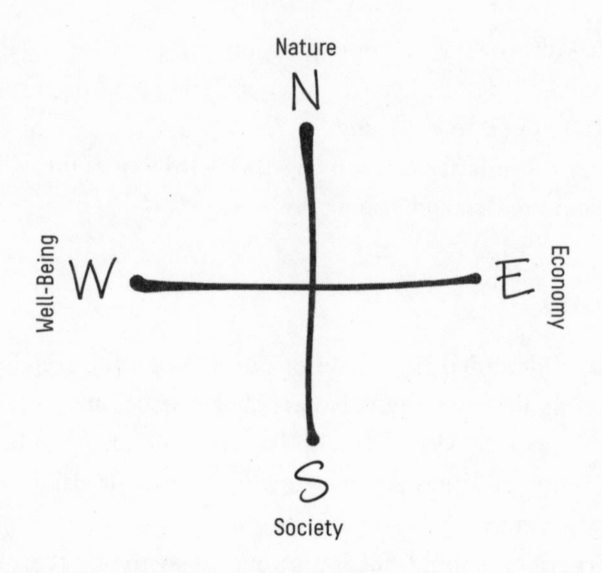

The Sustainability Compass.

the same familiar four letters, NESW, but we change the names: Nature, Economy, Society, Well-Being.

Using the Sustainability Compass ensures that you won't leave out anything important when you think about the issues and trends. The environment, economic, social, and individual health and well-being dimensions are all important.[7]

When we feel that we have a good-enough grasp of the relevant trends (it doesn't have to be perfect), we move on to the next step: systems analysis.

Systems

The word "system" simply means a set of separate elements that work together to make a whole and accomplish a specific purpose. Systems are all around us: examples include your body, your organization, and even the whole planet. Indicators help us to see the elements of the system we are working in. They are like pieces of a puzzle. The next step is to put the puzzle pieces together.

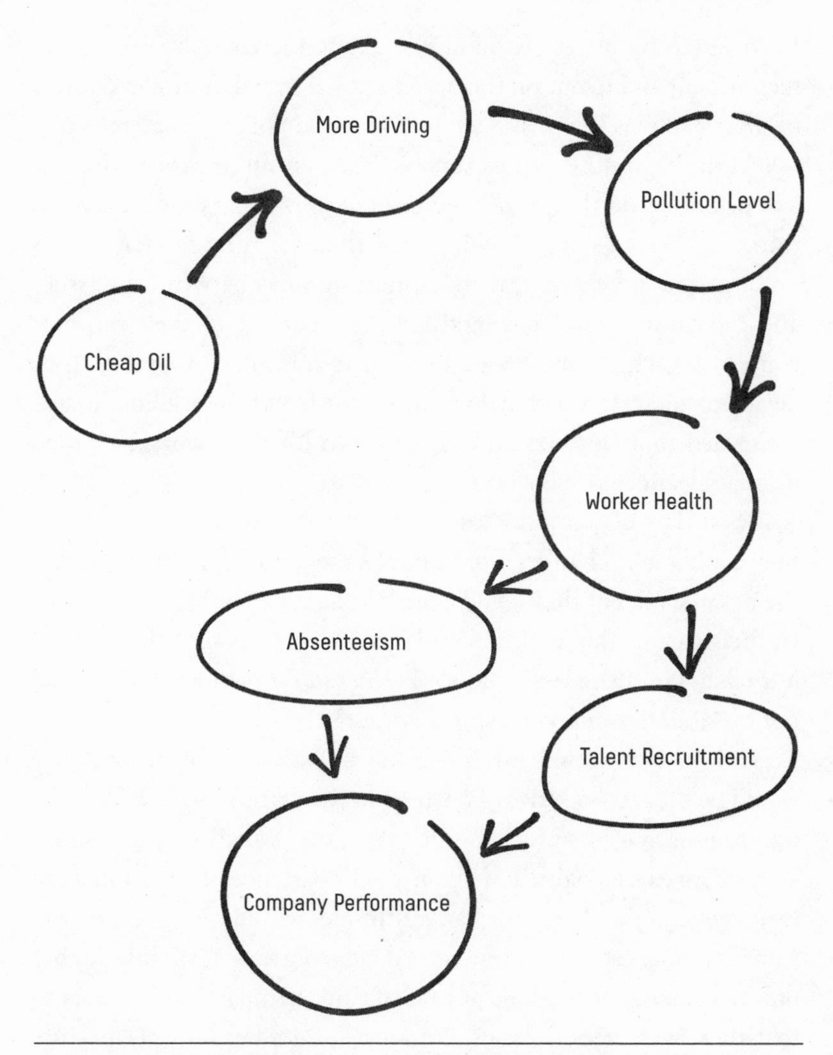

A Simple Systems Map. This systems map (or connections map), in its simplest form, shows the connections that an auto sales company made to understand how cheap oil could ultimately have a negative impact on company performance.

We were recently working with a large company in the automotive sales industry. They were looking for ways to become more sustainable, so we used VISIS to help them explore that. At the Systems stage, they began to see how some of the Indicators they had identified fit together: oil prices (Economy), pollution levels (Nature), worker recruitment and retention (Society), and the overall health and safety of their staff (Well-being) were part of a chain of cause and effect.

At the beginning of the exercise, they were all convinced that the low prices for oil that were then present in the market were a good thing for their business—selling cars. Then they drew systems maps on large pieces of paper, with connecting lines between the issues and indicators they had identified as important to their business context, showing how these and other trends linked together. They began to understand that low oil prices led to increased car use, which led to a worsened health situation for their workers, which ultimately affected their business performance.

Then they began to look for *where*, in that web of connections, they could make changes. For example, they could not directly affect the price of oil, but they could begin to change their business model. In their part of the world, hybrid electric vehicles (which consume less oil and produce less pollution) were not yet popular. So they had not even tried to market them very much.

Hybrid cars would not solve the pollution problem, but they would be a start to a series of changes in their whole market. In addition to reducing fossil fuel consumption, they would raise awareness about new technologies and about the importance of environmental issues generally and help to change attitudes within their very conservative car market. They even earned the company a slightly higher profit. Thinking about linkages helped this company to understand that there were good reasons, even just from a business perspective, to work harder on that. This is an example of "systems analysis" in its simplest form. You look at all the trends and issues that are affecting your situation, try to understand how they link together . . . and search for "ahas!" These are moments of insight about where to make change. Like this: "Aha! *That's* the area where we should focus our attention, that place where several issues are linked together in a chain. We've been treating the symptoms of our problem, which are over here, but not really looking at the root causes, which are over there."

Places like this in the system, where change can have multiple beneficial impacts, are called *leverage points*. "Leverage" means that a small effort *here* can multiply and spread, leading to even bigger change over *there* as well. When you spot a leverage point, you can see how changing something in the system, right there, will ripple out through the web of cause and effect, become amplified, and cause lots of other positive impacts. It's a good feeling!

Once you have found a number of leverage points, and had a few of these "aha!" moments, you are ready to move on to the next step.

An aside: Before moving on, we just want to point out that the topic of systems is very deep. You can spend a lifetime studying it. In a little book like this, we can only touch on systems and provide a few pointers for getting started. But we firmly believe that when doing systems thinking, every little bit helps . . . a lot. You don't need to be an expert. Even just drawing connections between things on a piece of paper can dramatically advance our understanding of complexity, so that we can see new ways of making change.

But if you want to start learning more about systems, we strongly recommend the classic primer by Donella Meadows (a famous figure in that field), Thinking in Systems.[8] *It will introduce some of the basic technical terms, but it also has a wonderful philosophical perspective on systems. As Meadows notes, "We know a tremendous amount about how the world works, but not nearly enough. Our knowledge is amazing; our ignorance even more so. We can improve our understanding, but we can't make it perfect."*

Innovation

Now we come to the part of the process that probably feels most familiar to change agents: the part where we identify a new idea to introduce into the system that we are trying to change. But it might be that we approach this part of the process in a slightly different way from what you are used to.

The word "innovation" is not the same as inventing something. When we are innovative, we introduce a change into the system, something that is not already there. And it can be a change of any kind: a process, a technology, a rule or policy, even a shift in attitude or culture. Innovations come in many flavors and sizes. In fact, they don't even have to be new. (Consider "organic agriculture," which people look at as an innovation, even though it is actually thousands of years old.)

The Systems stage helps us identify *where* we should be focusing our attention and making change. At the Innovation stage we identify *what* to do.

To conduct the Innovation phase in its most basic form, you can use simple brainstorming, and you can start with an inventory of existing ideas. What change ideas do we already know about, that might be appropriate ways to intervene in the system we've just mapped out?

Then it's time to think outside the box. Generate as many ideas for change as you can (this is usually easier to do with a group). Get them all on the table, even if the ideas seem a bit unrealistic or even crazy at first. Because you never know: from a systems perspective an idea might not seem so crazy after you study it for a while.

When working with that African country we mentioned before, helping to advance the idea of a green economy, we discovered that the actors in the government, the education system, and the private sector were not really talking to each other about such things as what skills the employers actually needed. In that system, information was not flowing where it needed to go. This was an "aha!" moment: we had found a leverage point.

We saw that by bringing these actors together, ideas would flow faster, information would spread, and the green economy process could accelerate. We knew *where* to make a change. But *what* innovation should we apply? Common ideas such as newsletters or conferences did not seem effective.

Then one of us (Axel) had an idea. It seemed a bit crazy at first, because no one had ever done it before: asking a well-positioned private sector company (a large accounting firm) to host a roundtable dialogue for representatives from all the key actors. This kind of meeting was usually hosted by government, or by development agencies. But few private sector people would usually attend; in fact, they were rarely even invited.

It took a little convincing, but soon enough all the stakeholders began to see the benefits of taking a new, fresh approach. The meeting happened—and everyone was very satisfied with the results.

This example of innovation also brings us to the next step in the VISIS method.

Strategy

Once you have an innovation that you want to apply, you need a strategy for effectively implementing it. In the case of our example above, the strategy was fairly simple: meet with the various representatives, describe the idea, and seek their agreement to participate in the meeting. Fortunately, the private sector company agreed to host the meeting and cover the costs as well.

But of course, some ideas will require a much more extensive approach to implementation. Strategy, like systems, is a very deep topic. You probably already know a lot about it, and there are many ways to improve your knowledge and skill on strategic planning. This kind of information is very commonly available, so we won't spend much time on it here.

Instead, we want to point out something else to you—something that you've probably begun to see for yourself. All the tools and methods we've been talking about (and many more that we're not talking about!) can work together. Here's an example of what we mean.

If you have a new idea, you can use the Amoeba model to map the actors in your social environment, so you know where to go first in introducing it and how to build up initial support.

If that idea is a complex one that needs a lot of buy in and a highly developed strategy to be implemented effectively, you can then use the tools of Stakeholder Dialogue. (Amoeba can even help you figure out whom to invite to which meeting.)

You could even use VISIS to structure one or more of those Stakeholder Dialogue meetings. (If you want, you can try using the pyramid workshop we mentioned earlier. Building a physical pyramid as you go through the steps really helps the group stay focused—and most people find the process enjoyable as well.)

And then, of course, when it comes to strategy and strategic planning, there are many other tools and methods you can apply.

Indeed, you can feel free to create your own recipe, mixing these tools and processes, or pieces of them. You can add visioning techniques, other formal indicator and measurement tools, systems mapping and modeling techniques.

This is the important point:

> A key skill for the change agent is the ability to look at all the tools, methods, processes, and techniques that are available, and then choose the right tool for the job.

Remember that quote from Maslow again, about hammers and nails: to be a good carpenter for change, you need more additional options than just a screwdriver and a saw. You need a whole toolbox.

If you're reading this book, you probably already have a number of tools that you use when you're trying to make change happen.

Which ones are your favorites? Why?

9

Focusing Change at the Right Level

KEY MESSAGES

- Change can happen at many different levels, from the surface to the deepest roots.
- No single level is better than another. Different types of change require action at different levels.
- The Theory U process developed by Otto Scharmer is another useful tool to add to your toolbox of change—especially for when you need to make change at deeper levels.

Here we are again, in our cozy little coffeehouse. It's a good thing we are feeling cozy, but in this part of our virtual meeting, we would like to start with a warning, and then conduct a little experiment.

The warning is that after our little conversation, you might need to undo all the preliminary work you've put into planning your change process, even if you've been using the stakeholder dialogue model or the VISIS method.

The reason is that you might discover that, yes, you have made a good change process plan . . . but it's targeted at the wrong level. Maybe you'll decide that you are still focusing on the superficial outer layers of the problem or challenge you're facing. Or maybe you'll discover that you've aimed *too* deep.

To illustrate how this could happen, imagine that you have been asked to plan the renovation process for an office building. You could approach your task by looking first at the colors they should use on the walls and floor, to mirror their corporate design or to make the office environment a little more friendly. Since people still need to work there, you would need to make a feasible schedule for the painting and carpeting work, to not disturb the office workflow too much.

But maybe at one point you notice that quite a few walls of the building are wet. The water is seeping in through underground channels after a heavy rain. You could still just paint everything, but you would know that this is only covering up the real problem and dealing with it on the superficial level. To go deeper, you might need to bring in an engineering team and fix the structural issues with the building. (But you still might confine the other changes to paint color and carpet.)

Or if you are in China, maybe you'd need to bring in a feng shui expert. You would see the renovation as a great opportunity to reorganize interior walls, furniture, and everything else according to feng shui principles. This would definitely mean working on a deeper level of functioning of the office building than mere paint and color—but it would still be focused on the building itself.

Or you might see a great opportunity to support a specific organizational culture change through renovating the office building. You might want to create a cost-efficient culture using an "open landscape," where nobody owns a specific desk or office. That way people can exchange news and ideas with each other and meet and brainstorm more spontaneously. You might add nicely designed "coffee spots" with creative tools and materials where people can meet, share, reflect, and brainstorm.

Another possibility might be to use the renovation project to support an organizational aspiration to integrate sustainability as a core guiding principle into the organizational DNA. The renovation project could involve digging into core principles, holding a stakeholder dialogue process, even integrating customers, investors, and representatives from the public sector and the civil society. You might look at the question, "How does a modern company live up to its responsibility to be sustainable? And how should that be reflected in our office building?"

All of these approaches (and we could have added a dozen others) are metaphors for different *levels* of change. One is not necessarily better than another, as they all serve a specific purpose.

But coming back to the start of our little journey, and the question about your own mental model, you will see that if you only have a hammer—or in this case, a paintbrush—then you would always advise the owner of the office building about the beauty of a newly painted office. But if you have tools for supporting other cultural change processes, plus tools to handle sustainability, then you could also explore with the owner the opportunity to move in the direction of becoming a truly sustainable organization. The renovation would be the outer, visible event that stimulates a deeper way of thinking, working together, and presenting a new profile to the public.

As we promised, we would like to add a little experiment into our conversation, which shows what we mean with the idea of "depth." Do you like ice cream? In our workshops we usually use a raisin for this, but here in this cafe, let's use ice cream. Order a dish of your favorite flavor. (If you don't like ice cream, just order something you like!)

Knowing that you have probably eaten this flavor of ice cream hundreds of times, we want you to slow down and taste the first bite with your fullest attention. We will guide you through this experiment step-by-step.

1. First, it might feel a little odd doing this experiment, and your mind may react with a lot of internal chatter. If that happens,

just notice any thoughts that come up without paying too much attention to them.

2. Now just look at the ice cream for a moment. What do you see? What shades of color are there? What is the surface structure of your ice cream?
3. Next, with your full attention, take your spoon. When you touch it, how does it feel? How cold or warm is the spoon? Feel the weight of the spoon when you lift it and move it to the ice cream.
4. How easily does the spoon move into the texture of the ice cream? How does the weight of the spoon change when you lift it, with the first bite of ice cream, toward your mouth?
5. Be very sensitive to the moment when you first smell the fragrance of your favorite ice cream. Also be alert to when you first notice the coldness of it, as it moves toward your mouth.
6. How does it feel when the ice cream touches your lips? When you feel the temperature and the texture, and you taste the taste?
7. What happens when you allow the ice cream to slowly melt in your mouth? What sensations can you notice? What happens when you swallow? All these sensual experiences might trigger some thoughts, memories, inner pictures, inner sounds, feelings, or body sensations.

How was that little experiment for you? Did you feel any different afterward?

Usually, in our live-training sessions, people say that they feel calmer and more relaxed after this little exercise and that they were astonished at the newfound intensity of tasting something they'd eaten hundreds of times before. When we do this in a group setting, the whole atmosphere changes, and there is a moment of quietness and subtle happiness in the room.

You might be a little puzzled and ask yourself why we would do this kind of experiment to reflect on improving our effectiveness as change agents. For us this little exercise is a pointer toward one of the biggest leverage points for creating successful change processes.

Remember that in the beginning of our little journey, we quoted from a McKinsey report that said only about 30 percent of all change processes are successful. We believe that one of the main reasons for this lack of success is not because people don't have good tools and methods, or because change is next to impossible. We believe that change often fails because the people involved in the change process have forgotten how to "eat ice cream" in the way we just described: very attentively.

To explain what we mean, let's take a little detour into the field of neuroscience.

Every single experience we have is processed in our brain by stimulating neurons to communicate with each other. The more often we have those experiences, the more efficient and stable become the connections among neurons. The following image is often used to explain this mechanism: Imagine a landscape. A person walks across it. A moment later you might be able to see some marks in the ground indicating where she walked, but a day later no marks would be visible. But if that person walks the same path a couple of times every day, you'll see that a little path has formed. And if more and more people use that path, it becomes larger and clearer. Maybe later it will be used so often that it makes sense to build a road or even a highway.

The connections in our brain that are used the most are like neuron highways, where communication is very fast and very predictable. We have paths, little roads, and even high-speed highways for everything we experience, whether it be for finding the right words to say what we mean, for letting our fingers move in the right way to play a piece of music on the piano—or for observing what is happening in a change process.

If we usually have the mental model, "The glass is half empty," then we will have very well-developed neural pathways for seeing things that way. Our mental models are not more or less than patterns in the connections of different clusters of neurons in our brain.

The brain is the biggest user of energy in our body. Although it only accounts for 2 percent of the whole body's weight, it uses 20 percent of its energy. To use that energy as efficiently as possible, it builds these kinds of patterns, which allow it to run most activities in a kind of automatic pilot mode.

Axel: I still remember vividly how exhausted I felt after my first long-haul driving experience, when I had just gotten my driver's license. I had to be fully conscious about almost every action I took, such as changing lanes, passing another car, and so on. I had no built-in routines for these actions. It took some time before the first stable paths were generated in my neural network, which allowed me to do these kinds of things without thinking about them anymore. Now I don't need a lot of energy to drive the same distance, because I don't have to think very hard about it. I have become much more energy efficient by doing the same things in a habitual way.

Some researchers say that about 90 percent of all our actions are done in this automatic way. That's very good for energy efficiency, but it is not at all good for creating something new.

The neural patterns in our brain structure the routine procedures not only for driving but also for eating ice cream. Sometimes we've finished our lovely ice cream without having really noticed how it tasted. This happens, for example, when we eat ice cream and read an interesting article at the same time, or talk to a friend, or rush to a meeting. We can do these things in parallel, because we can trust the automatic pilot in our brains.

But as a consequence of using our automatic pilot, we also often come up with habitual ideas and proposals when facing a challenge in a change process. We tend to approach the challenge in the habitual way, which has proven appropriate in the past—so we take out the usual hammer.

You find these habit patterns not only in individual brains but also in the culture of a country, an organization, or any other grouping of people. Consider the habitual way that the concept of "economic growth" is always the default strategy when thinking about the progress of a nation. Or the way finance people often think first about controlling costs, whereas marketing people look at advertising, and sales people ask for higher incentives for the sales reps, to keep the organization performing at its best in difficult times.

One of the most important functions you have as a change agent is to help the people in the target audience move beyond their habitual ways of approaching the issue. At the same time, you also need to distance *yourself* from your own habitually patterned thoughts, ideas, concepts and actions.

In other words, you need to have the courage to "eat ice cream" in the way we described earlier with members of the executive board or the steering committee or the project team or others . . . to help them move beyond their habitual ways of responding to the given challenge.

To help people navigate this process, Otto Scharmer, an MIT-based expert on cross-sector innovation, developed a framework called Theory U.[9]

If you really want to help the situation change in a deeper way, you need to help the people involved focus on a deeper level of reality. The U-shape referred to in Theory U is a metaphor for going deep, getting past our own blind spots by learning how to see our situations in new ways (as we did when tasting the ice cream), then coming out somewhere new. Scharmer calls this "presencing."

Scharmer describes several stages in this process.

Downloading Past Patterns

At this stage you, as a change agent, need to persuade the stakeholders involved to try something new—something that is often beyond their comfort zone. So the stages along the U's downslope

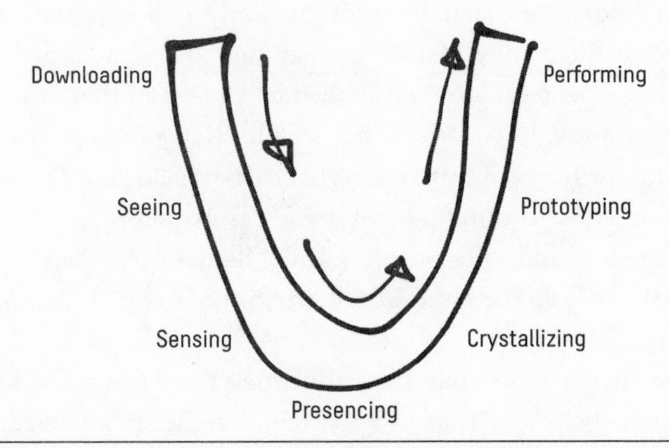

Otto Scharmer's Theory U. This simplified adaptation of Otto Scharmer's Theory U shows the stages of presencing, a process during which people involved in a change process move down one side of the U, awakening to the current situation, and up the other side, where one begins to see a new possible future. This version was adapted from Wikimedia.org, an open source Internet archive. If you want the full exposition of the theory and how to use it, you should see Scharmer's book, *Theory U.*

are aimed at suspending our automatic reactions, which Scharmer calls "downloading," and setting them aside.

The more we feel ourselves to be under pressure, the more likely we will respond to a challenge with our habitual old answers—essentially downloading our old behaviors. If we do this attentively, we begin to understand our own automatic pilot modes. You can't change old habits without first recognizing them.

Seeing with Fresh Eyes

Do you remember how we helped you experience your favorite ice cream with fresh eyes by helping you focus all your attention on that one sensual experience? In a similar way you need to support the stakeholders to look at the given reality—including the available data and their own observations—without using their habitual patterns. You need to help them see new details, new cause-and-effect relationships, and often an expanded view that broadens and deepens their field of vision. (In our work, for example, this is when we bring our Sustainability Compass into play, or move to Indicators and Systems in the VISIS process.)

Sensing from the Field

You are part of a broader stakeholder group. How does reality appear to *other* stakeholders? "Sensing from the field" means getting a sense of how others perceive the situation. If you are a business owner, how would this reality look though the eyes of one of your workers, or your suppliers, or to a person from Greenpeace? The list of vantage points should be long.

When everybody in a process begins stepping into the shoes of others, their individual perspective starts flowing into the broader sense of the whole, or "the field." But Scharmer also notes that this group perspective is still a result of the past: for a different future to emerge, you also need to let that go.

Presencing—Connecting to Source

Presencing involves sitting with the deeper questions about what we are doing. What is our real purpose? What is the meaning of this work? What do I want to be remembered for when I die? Most of us have an understanding that life is about more than accumulating wealth or achieving personal goals. Before we move back up the other side of the U, it's important to stop and connect to the source of our deepest values—whatever that source is for us—and allow ourselves to be surprised at what wants to come up.

Crystalizing Our Vision and Intention

The moment of presencing often is a very quiet moment, when the habitual chatter of the mind is at rest. This moment of mental and physical rest in the brain allows for more parts of our self to get involved in the process of visualizing or formulating something new. Sometimes the process leads to a new feeling, or an image, or even a melody. It can bring up a phrase that seems to appear from nowhere. When these creative products of deep reflection are shared in a group, they can trigger other images, phrases, or feelings in the minds of others. Slowly a collective idea is born, one that no individual can claim as his or her own: it is just there, and the time was right to perceive it.

When the new vision has become clear enough, it is time to move up to the phases that involve action.

Axel: I want to take this moment to share some interesting neuroscience about the "aha!" experience—that moment when a new vision or intention is crystallized. According to cognition researchers John Kounios and Mark Beeman, a burst of high-frequency gamma waves (40 Hertz), a sign of very high brain activity, can be detected in specific parts of the brain at the moment of insight. But interestingly, the moment before that burst happens, the brain activates an alpha-wave state (approximately 10 Hertz).[10]

These alpha waves effectively block the processing of other information in the brain, so that the insight or "aha!" can come into awareness. This moment, for me, is similar to the moment of tasting your favorite ice cream with full focus and attention, which means some other information has to be blocked out for that moment.

Prototyping the New

While keeping that feeling of deep connection with the whole of oneself, and with the group, the new idea needs to be given form and shaped into a kind of prototype. What will this idea—an umbrella, really, for your change process, initiative, project, or other endeavor—be called? How will it work? A prototype is a kind of first draft of the eventual outcome that lets you test the idea and reshape it if necessary, taking care not to lose its newness and originality by a return to those habitual old ways of thinking.

Then you create a *new* habit, establish a new set of neural connections, by using the prototype.

Performing by Operating from the Whole

A new idea has been born out of the intelligence of all the stakeholders involved. Since the knowledge, experience, and needs of everybody involved has flowed into the creation process, it creates a solid foundation for sustainable performance. People will continue "enacting" the idea, based on their common experience of having interrupted their usual patterns, reflecting deeply, and sharing the results of those reflections. They will keep the whole in mind, as they take care of their individual role in the new reality.

* * * * *

We're sure that you are continuing to notice that there are similarities between some of the tools and methods we're describing here, as well as some obvious points of connection, where one tool can support another.

Some of the stages in the VISIS Method, for example, are a bit parallel to those of Theory U. The systems-thinking stage, for instance, is also a point where we work to see with fresh eyes. And the ideas and concepts in Amoeba can be very useful if you're trying to convince a group to try a new process (such as Theory U, VISIS, or the Dialogic Change Model) in the first place.

We want to encourage you to see such connections and to approach all of these tools the way a good carpenter approaches her whole toolbox. There is not just one kind of wrench: there are several wrenches, because different kinds of wrenches work better in different situations. And sometimes you have to use a special combination of tools to get that unique job done.

Change tools are also like this. We're not trying to give you a new hammer! No single tool will always fit in every situation. For example, we sometimes use ideas from Theory U in our work, but not always, and we usually combine them with other ideas, too.

You can even apply a process such as Theory U to yourself and use it to examine your own approach to the change process, as a way of helping you perceive the situation freshly and to figure out the best approach to working with a group. It might lead you to using Theory U, or it might lead you to using Appreciative Inquiry or VISIS or Amoeba or another tool or method—or some special combination.

That's entirely up to you.

So now, a question to you: what has this chapter made you think about?

10

Increasing Your Personal Impact

KEY MESSAGES

- Leading a change process often involves working without the usual tools of power, such as having a hierarchical position (where people have to do as you say) or a large budget.
- In those situations you have to be persuasive and convince people to follow your recommendations, suggestions, or requests.
- You can improve your chances of success by paying attention to three concepts: Authority, Presence, and Impact (API).

As a change agent you will often find yourself in a kind of leadership position—where you are trying to make something happen—but without having any of the real insignia of power, such as an important title or a large budget. In those situations no one has to follow your orders: you have to *convince* people to do as you suggest.

Or you might have been put into that position by people who expect something from you: after your engagement *something* is supposed to have changed. Sometimes the goal of the change process can be very obvious, such as having a renovated or greener office building, or even installing a new computer program. But sometimes it's a little less visible, such as creating a new approach to talent management. And sometimes the main target is completely immaterial, such as changing people's mind-sets.

Axel: Once I was part of a team leading an immaterial change process of this kind. We were asked to help the European branch of one of the top global companies for printing and copying machines to become truly Europe-focused and to change from the mind-set of "installing machines" to "providing document management solutions."

This was a big challenge for them, as well as for us. The company representatives were real techies, proud of the speed and resolution of their machines. Internally, they were called the box movers—because they sold and installed big machines that came in large boxes—and they enjoyed talking with the techies on the customer side.

But now they had to build good relationships with office managers, for whom technical details about the machines were not really relevant. They had to start talking about how the machines would fit into a new, expanded document management process. They had to learn a new language and new sales techniques and to adopt a whole new mind-set regarding what their business was actually about.

Our job was to lead the change process, but we were not executive officers in the company. We were the consultants. Here's how we did it.

First, we created sets of peer groups, composed of box movers. Everybody in all the groups was informed about the need for this strategic change, and they all agreed to the reasons driving it. The necessary shift they needed to make was on the level of self- identification with the new role and on the level of behavior. In the peer groups, participants could reflect together very openly about what actions they had tried to make these shifts, which ones had worked, and what challenges they encountered. Through these groups the attitude of the box movers changed quite dramatically from a reluctant, "I don't want to, but I have to" to a positive and curious, "We'll make it."

Being in this kind of situation is not necessarily a weaker position. A leader with the insignia of power might have the assertiveness to force people to follow his or her instructions, but the official power does not translate automatically into a sense of conviction within the target group. Leadership of this kind does not always create *followership*.

Followership—the conviction that the goal is right, combined with a commitment to support the process of getting there—is always in the hands of the followers. It is given by free will.

To create followership, it doesn't make much difference if you have the insignia of power or not. Look at great historical leaders such as Mahatma Gandhi, Martin Luther King Jr., Nelson Mandela, or Aung San Suu Kyi. None of them was in a position of formal, official power when they created movements that changed nations. But they were perceived by large numbers of people as leaders whom they were willing to follow into the most challenging situations. Real leadership is like a dance: it always needs both leadership *and* followership.

One of the most fundamental conditions determining whether people will follow you is trust. People will not follow your lead freely unless they feel assured and confident about you as a person—no matter whether the change process you are leading is a small office-related matter or a large societal change.

Please take a sip of your coffee or tea, and let's take a moment to reflect on the qualities you might need, so that people might give you the benefit of their trust.

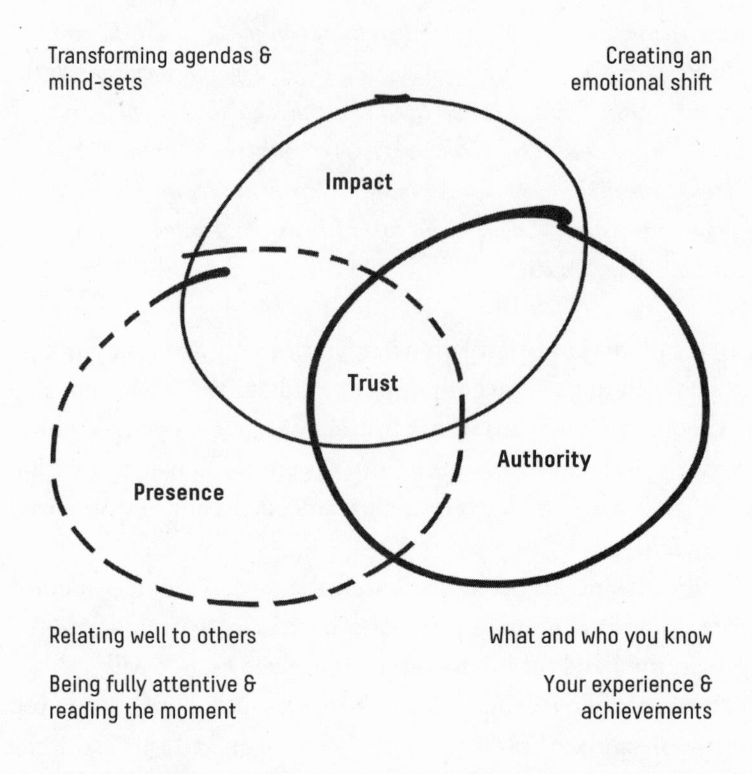

The API Model. A representation of Peter Hawkins's API model of how to build trust and influence in a change process. (You can learn more about this diagram in the main text.) Source material courtesy of Peter Hawkins.

When we talk about trust in this context, we're not using the word in the sense of "reliable," as when we are talking about a trustworthy person. Of course, telling the truth and doing what you say you will do are fundamental to trust of any kind. But we're talking about trust as a kind of *default attitude* that you have toward a person. That trust makes you more likely to follow her advice, suggestions, and requests. It is a kind of *feeling* that we have about people. Where does that feeling come from?

Some years ago a good friend and colleague of Axel's, Peter Hawkins from Bath Consultancy Group, developed a little model for one of his clients, a big multinational accounting company.

They were looking for ways to help identify who among the junior staff might be good candidates to advance into the leadership levels of the organization. The model is called API, which stands for Authority, Presence, and Impact. We'll walk you through these three ideas now.

Authority

If you needed to get knee surgery, you might immediately start searching for the best surgeon. If you found somebody who was performing such surgeries for the best athletes, your trust in him or her would definitely start to grow. If the surgeon also held a professorship in a well-known medical university, and led the best clinic in the country, you would be very likely to choose that person. The surgeon would have, in your eyes, *Authority*.

Your own Authority is connected to your past. It is composed of what you have learned and achieved but also whom you know and what kind of title or position you have.

In addition to these outer aspects, Authority also has an inner aspect to it, which is connected to your own belief in yourself. If you have a lot of self-confidence in what you can bring to the table, you will radiate a number of subtle, nonverbal signals that can be interpreted easily by others. The process is rather unconscious, which is why we often use language such as this: "She has an aura of authority about her." This nonverbal aura that you create—which, in reality, is a pattern of behaviors—is often more important to your success in convincing people to follow your advice than is the deepest knowledge and experience you might have.

When your knowledge, expertise, competency, and experience have built the foundation for a solid feeling of self-esteem and confidence inside you, this is perceived by others as your Authority.

Please take a moment and think about two or three people you know, or have seen in the public eye, that you would say have great Authority.

- What are some of the things you notice about them that give them their sense of Authority?
- In what way would their Authority influence you, if you were to interact with them personally?
- Please imagine an "Authority scale" starting with 0 (= no Authority at all) and ending with 10 (= highest possible Authority). Where on that scale would you put the people you have just been thinking about?
- Where would you put yourself on that scale, in your role as a change agent in a real change process (or anywhere else)?
- What is it exactly you are building on as your Authority? Is it a title, your expertise and knowledge, your experience and age, or other factors?
- Is the sense of Authority that you signal to others something you believe you should work on, or are you content with it?
- If you feel you should work on your sense of Authority, do you have an idea of how to go about it? Please remember that very often it is not so much the "hard facts" that determine whether you are perceived as having Authority but your own self-esteem and confidence.

Presence

Here are two very similar stories that Peter Hawkins often shares in his presentations. They illustrate the concept of Presence (quite different from the "presencing" process we explored in the last chapter), featuring two very different leaders: Bill Clinton and the Dalai Lama.

Sometime after his presidency, Bill Clinton was leaving a meeting and went to the garage to wait for the car that was coming to get him. In the garage many drivers were there waiting for their "famous" bosses. As Bill Clinton passed by he shook their hands and talked for a moment with each driver, asking who they were driving and sharing reflections about those people. Afterward, each driver reported feeling that, at that moment when Bill Clinton was talking to them, he was really focused on them. They felt that they were having a real and important conversation with him, even though it was very brief.

Similarly, the Dalai Lama was once invited to a dinner with officials of the European Union. On his way to the banquet, he passed by the kitchen. Spontaneously he sneaked in and started chatting warmly with the kitchen staff. Afterward they all said they had the impression that for a moment they were very close emotionally to the Dalai Lama.

Bill Clinton and the Dalai Lama seem to have the skill of *Presence*. They can relate to people, even strangers, in a way that makes those strangers feel seen, listened to, and appreciated. They create a feeling of being "in the moment."

These are famous examples, but Presence is something everyone can develop. If you have Presence, people enjoy talking with you. They more easily open up in front of you, because they perceive that you are interested, and that you understand them.

As we mentioned before, we might trust the best expert surgeon for a complicated knee surgery. But when we go looking for a family doctor, the ability of that person to listen well and deeply might be more important than any fancy academic titles or whether any famous patients go there.

Please take a moment and think about two or three people you know, or have seen in the media, that you would say have great Presence.

- What do you notice about them, and their Presence?
- In what way would their Presence influence you, if you were to interact with them personally?
- Imagine a Presence scale, like the Authority scale, from 0 to 10. Where on that scale would you put the people you have been thinking about?

- Where would you put yourself on that scale, in your role as a change agent in a real change process (or anywhere else)?
- What is it exactly you are doing when you are being present? How do other people notice your Presence?
- Is your skill at being present, and being felt as present by others, something you believe you should work on, or are you content with it?
- If you feel you should work on your Presence, do you have an idea of how to go about it?

Here is an important note: it is almost impossible to be present with others if you are not present with yourself. So in general, learning the skill of Presence starts with building your own capacity to stay mindful about your own experience of the here and now, without your thoughts becoming trapped in reflecting about the past or planning for the future. "Eating ice cream," in the way we described in the last chapter, can be a key starting point to establish your own Presence, and to be perceived as Present by others, which will in turn build trust.

Impact

The success of change agents is ultimately measured by their ability to convince people to think, feel, and act in a different manner—their *Impact*. Impact is the brutal final grade, determined by whether or not your actions bring the desired results. As change agents we can never be truly satisfied when the testimonial we get is, "Well, at least they tried."

Impact does not always have to be the result of assertiveness, or even Authority. Direct actions and statements can obviously produce Impact, but so can good listening; that is, Presence.

Alan: An old friend and colleague of mine, Duane Fickeisen, was once asked to help a government water program involving a lot of stakeholders. (Duane was a consultant at the time, but he later became a minister in the Unitarian Universalist church.) The stakeholders' change process was in trouble. They wanted Duane to facilitate and help them iron out their conflicts and difficulties. But first, he just observed one of their meetings. It seemed to go rather well.

After the meeting several people came up and thanked him for being there. They believed he had made a positive difference on their meeting, even though he had not said a word. Somehow, his careful, appreciative observation—his Presence—had made an impact.

Here are some other examples. Suppose you are holding a meeting with a lot of critical and concerned stakeholders. After affirming that you have heard and understand their concerns, you convince them to "park" or suspend their criticism for a time and at least listen to your arguments.

Or imagine a meeting with members of your change team who are frustrated because things are not progressing as planned. You find the right words, and those team members leave the meeting energized and with more optimism.

But the ultimate Impact, in most change initiatives, is a cycle of actions and behaviors that produce some measurable results. Thanks to your initiative, people actually enact the change.

As hinted above, Impact is connected to Authority and Presence in a couple of ways. On one level it stands next to them (see the diagram). It is technically possible to have Impact even if you have relatively low levels of perceived Authority and Presence, so it is possible to separate these three. A persistent and clever activist might make an Impact on a company, for example, even without being perceived as having much Authority or Presence.

But for change agents these three are usually combined. We have put "Trust" in the middle, where the three qualities meet. Our experience is that people trust you more when they feel that you can have a positive Impact on them. If I notice that after a meeting with you, I feel better or think differently, or I notice a real change in my behavior, my level of trust in you increases.

But if you look closer at how to create an Impact, you might see that it is usually created when there is the foundation of Presence and Authority. If I trust your knowledge and expertise, and at the same time I feel deeply understood by you, then the likelihood is very high that a good intervention from you might shake me up and create a shift in the way I think or act.

So besides creating a good relationship to somebody, building on your Authority and Presence, you also need to know how to make a good intervention that creates an Impact. To find those intervention points in the other person, the group, or the system as a whole, you as a change agent often need to be spontaneous, innovative, and courageous. Impact tends to happen when something is done unexpectedly. You can't create something new if you continue doing the same thing again and again.

But on the other hand, after being created, an impact or shift in the system needs to be stabilized and habitualized. And this only happens by doing it again and again.

Take a moment to reflect about two or three people you know, or have observed, who appear to create a lot of Impact.

- What do you notice about their ability to have an Impact? What are they actually doing?
- In what way do they—or would they, if you talked with them directly—influence you with those activities?
- Imagine an "Impact scale," like the previous ones, from 0 to 10. Where would you put the people you were thinking about?
- Where would you put yourself on that scale, as a change agent in your current change process (or in other contexts)?
- What exactly do you do to create Impact? How are other people noticing your ability to create Impact?

- Is your level of Impact something you believe you should work on, or are you content with it?
- If you think you should work on it, how could you go about it?

To get more clarity about your strengths and learning areas when it comes to being a leader that others are open to following, you could ask a couple of peers and colleagues to give you feedback using this API model. Just explain the terms to them, and then ask them for an honest opinion, or a "scale rating" like the one you gave yourself.

This is also a good exercise to use with the people you are working with in a change team. For example, there are situations where it is best to send somebody to a meeting who is very strong in Presence, while other situations might call for more Authority, to create a real shift in thinking, feeling, or acting.

*****

How did that feel, to take a look at your own leadership strengths and style, when it comes to change work? Was the API model useful to you?

How do you think you could use this model yourself, going forward?

11

Supporting Others to Perform Well

KEY MESSAGES

- Making change happen requires much more than just providing people with information and a reason to change.
- The concept and practice of coaching is a powerful way to improve your communication skills—and help others to improve their own skills, as well as their readiness to change behavior.
- While coaching is a profession, it is also a basic skill that any change agent can (and should) learn.
- Organizations can create a "coaching culture" to improve communication and the skill of change, in every context.

In our last conversation we focused on you. This time we're going to focus on other people. Specifically, we're going to look at how *you* can support *them*—so that they are able to support a successful change.

The success of any change process is measured by whether it creates the intended result. But on the way to creating that result, there are two different aspects we need to consider. On the one side, there is the *rationale* for the change, the reason for initiating it. On the other, there is the actual changed *behavior*, which involves people doing things differently.

We believe that one of the main reasons for not achieving success lies in the fact that change agents have a tendency to focus more on the correctness of the *rationale*, rather than the question of how to support people in adopting new *behavior*.

Years ago a study was done with patients who needed to undergo coronary bypass surgery, because the arteries supplying blood to their hearts were blocked. The reason for the blockage was that those patients were not exercising, not eating well, or otherwise increasing their heart disease risk over a long period of time. The main cause of their disease was bad habits, which could theoretically be changed. The patients were told that if they did not change their behavior, the likelihood that their arteries would become blocked again was very high, and the risk of having a heart attack would rise tremendously. So what do you believe? How many patients in that study group seriously changed their habits?

It was only 10 percent. Despite the real urgency, and the presence of the strongest possible threat to their lives, the strength of their habitual behavior kept 90 percent of them on the old, unhealthy path.

What improved the success rate? Giving the patients more information—strengthening the *rationale* for change—did not do it. But when the patients were invited to join support groups, where they learned new ways of healthy cooking, or exercised together, the success rate for changing their behavior went from 10 percent to 77 percent—as Alan Deutschman pointed out in a *Fast Company* article.[11]

Studies like this show us that change is possible. But it doesn't come automatically, just by identifying a strong need, or even by supplying a vision. It involves processes that engage people and a social environment that supports change.

The crucial question for creating such environments is always, "How do we create an impact? How do we win people over to the new way?"

The Crucial Role of Communication

If you are involved in a more complex change process, you will need the direct assistance of others to drive change forward. Maybe you'll need a dedicated change team. Maybe you'll need the CEO of the company to give a talk, explaining that this change is really important for the future success of the company. Or maybe you'll need an external expert to run a specific training or inspiration event.

Very often there are many separate interventions, by different actors, that will determine whether the change process gets another push and keeps moving in the right direction, or stalls. So it is important that you are able not just to convince people generally but to win over your change team members, the CEO, the external experts you need, and so on. (Remember the Amoeba in chapter 6: you need to recruit other Change Agents and the Transformers to your cause.)

With this in mind, we can see that the success of a change process depends on creating shifts in the minds and hearts of many people—and then helping them create shifts in the minds of many more, to drive the change forward.

We use the phrase "shifts in the mind" because we want to emphasize that the process of communicating to people about change, and convincing to them to actually behave differently, has many layers to it.

Communication passes through many steps before it results in action—and ultimately becomes a new habit. Simply "letting people know" is far, far from enough. First, they have to *hear* you. Even this is not automatic, since they may nod and say yes but have not really heard because they were thinking about something else. Then comes the question of whether they *understood* what you are suggesting. Understanding is still not the same as *agreeing*, and agreeing to do something is not the same as *trying it out* once or twice. Finally, just trying an idea is not the same as making it a

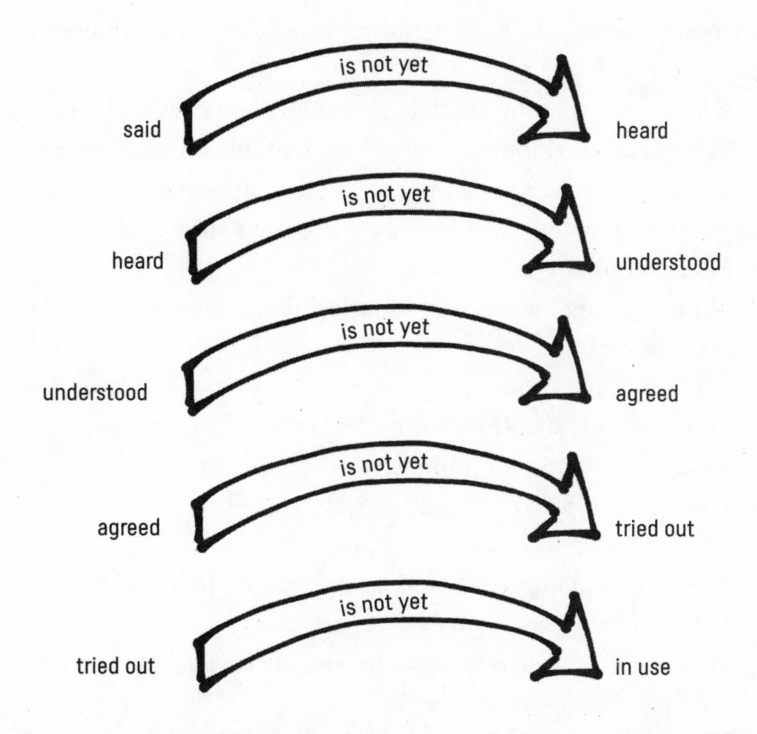

Steps of Communication. The steps from voicing an idea to implementing it are many. Based on an idea by Konrad Lorenz.

habitual part of a new routine; in other words, truly putting the idea into use.

The Biggest Leverage Point of All

So while content, stakeholder process, planning, and other elements are important to a change process, communication is central. This means that the biggest "leverage point"—the place in the system that you can influence to accelerate change and make change processes more likely to succeed—is you.

Obviously, sharpening your own communication skills is always a good idea. But even this is not enough. We believe that an even more subtle and effective approach is to work on how you coach *other* people, to help them improve *their* skills and perform at their best.

Bringing a coaching attitude to a change process, and building a *coaching culture* among everyone involved, means creating many moments where people simply talk—and listen—in a reflective way, with the intention of helping each other solve problems and improve performance.

Typical focus questions for coaching conversations might include these:

- How am I personally doing, as the leader of this change process? Where can I improve?
- How are you doing, in your role? How can I support you to improve?
- How are we doing as a team? Where can we improve?
- What can we learn from each other?
- Where are we falling into our habitual behaviors? And where are we losing our creativity?
- Which topics do I, or you, or we tend to avoid addressing?
- How can we come up with new creative options to address our challenges?

The more the success of a change process depends on *people* changing, the more this kind of coaching attitude should be woven into the DNA of the change team. A coaching attitude tries to leverage the unused potential in every individual on the change team, and in the team as a whole.

There are professional training programs in coaching that are up to three years long. Even in those trainings, team coaching is often not part of the curriculum. So please don't expect that after our next couple of cappuccinos in our virtual café, you will have become a professional coach, certified to work with individuals and teams.

But you can certainly adopt a coaching *attitude*, practice coaching *skills*, and start building a coaching *culture* within your team.

Indeed, we believe that this kind of communicative learning environment is crucial for creating success in any challenging change process. You don't have to be a certified professional to do that!

In a coaching culture you create a process of reflecting on *how* you do things, which is running parallel to the process of actually *doing* things. This parallel process needs to be rooted in a basic understanding of how we human beings are—to a very high degree—driven by patterned behaviors (habits) and by mental models developed through past experience. Many of these behaviors and mental models are very effective, but others limit our ability to have an impact on others.

To help each other to overcome limiting behaviors and mental models is one of the most important success factors in running a change process.

Building a Coaching Culture: The Basics

A coaching culture involves opportunities for one-to-one conversations, as well as group processes. Both need to be rooted in an atmosphere of mutual trust and support for each other.

Let's take a look at them both.

The One-to-One Approach to Coaching

As a change agent, you'll have to interact with your colleagues in many different ways. Sometimes you need to inform them about the latest developments. Sometimes you need to push them to deliver. And sometimes you might need to support them to improve their performance and increase their impact. Let's call that last kind of conversation coaching, with you as a coach and the other person as the coachee.

Generally, coaching will only be effective if the coachee *wants* to be coached, which means that he or she seeks some support by you. The coachee wants to reflect on an issue with somebody who can provide an outsider's view.

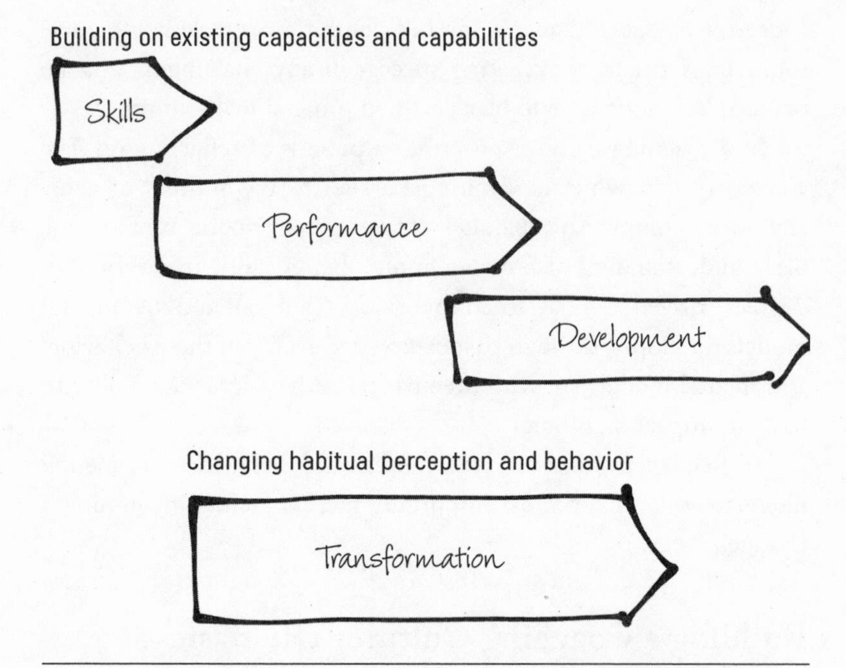

Essential Elements of Skill Building. This model was originally developed by Peter Hawkins and Nick Smith as a spectrum of activity, to show coaches where to focus their intent and effort. You can help the coachee develop skills, enhance performance, and boost development—or transform. We have emphasized this difference by putting skills, performance, and development on one level and transformation on a separate, deeper level. We want to highlight that transformational coaching, which involves making a significant shift away from preexisting patterns of thought and behavior, is a very different undertaking compared to the other three levels, which retain the coachee's current patterns but build on them. Based on a chart by Peter Hawkins and Nick Smith.

There are different types of coaching conversations, which go to different levels within the person being coached, as illustrated by the coaching spectrum, developed by Peter Hawkins and Nick Smith.[12]

Skills

Sometimes you and the coachee notice that specific skills are missing. These might be technical skills, such as developing sustainability indicators or writing a sustainability report. Or it might involve presentation skills, or the skill of planning and running a stakeholder dialogue process. With skills coaching, your main focus is to identify the skills needed; provide or point to the necessary tools the person might need; and allow (or encourage) the coachee to practice with them. It's also helpful to provide him or her with feedback once in a while about the progress you see.

Performance

A good performance is the result of a whole bundle of skills and activities, paired with an *attitude*. To help somebody increase her performance, the coaching process needs to look beyond what that person is doing, to what she is thinking and feeling while she does it. Is there the willingness and desire to be seen as a high performer? Is there any nervousness or discomfort, and are those feelings getting in the way? Is she doing the right things at the right moment, with a good sense of timing? To focus on increasing performance, you as coach, together with the coachee, need to really analyze the different factors leading to good performance, and work on them.

If you remember the very beginning of our little journey, you might recall our conversation about theories of change. You and your coachee's belief system about how humans change will determine how you approach this issue of performance, as well as how you approach the following two issues.

Development

Development is a gradual process of increasing one's competence and altering specific routines to match new situations. It requires the opportunity (and willingness) to *practice*.

For example, if you want to help a technical expert on your change team become the leader of a small subteam, you need to understand the development steps he or she needs to make. An expert often needs to have the ability to dig deep into a content issue (and many technical experts love to do that). But a team leader often needs only to have a basic knowledge of the technical details—and in fact should probably avoid digging too deeply into them.

Success in terms of the new responsibility shifts from technical mastery to providing a good working atmosphere for the other staff. If there is a general wish to grow into this new responsibility, development coaching is usually focused on how to shift from one way of thinking (and set of habits) to another. The coaching session can become something like a little practice session: "How do you imagine you would handle this new situation?" Specific issues might require a shift into skills coaching as well.

Transformation

When the focus is on transformation—a more fundamental shift in belief systems and behaviors—coaching moves to a different level. Transformational coaching is relevant when the coachee's old, underlying belief system no longer matches the needs of a new reality. A real transformation happens when a person who is used to looking at the world through the belief system "the glass is half empty" starts seeing things as half full. To support such a transformation, a real shift needs to be initiated inside the person. (Doing this kind of coaching effectively usually requires professional training.)

Basic Principles and Tools for Coaching

At one end of the spectrum, coaching is a real art, and mastering that art happens through serious training and then is fine-tuned over many years of experience. At the other end, even a good and open conversation with a friend in a café can have an effect similar to that of a professional coaching session. But this café coaching mostly happens by chance. As a change agent you need something more reliable than chance: your own intention to be a good coach.

Even at the beginner or "café" level, effective coaching follows the same basic principles and tools as professional coaching. By adopting these you can improve your own skills at helping others to improve theirs.

Parking your personal agendas: A coaching process happens more easily if one person is ready to "park" his personal agendas, to set them completely aside, and allow his full attention to be with the other person.

A relationship based on mutual trust: Trust is the fundamental base for opening up, sharing the real stories, and being ready to critically reflect on oneself.

Listening: This means not just taking in the content of what somebody is saying but also listening to the subtle "undercurrents" that accompany that content. These undercurrents can be expressions of the body, a change of the tone of voice, the pace of the conversation, an indication (perhaps signaled with a hesitation or a small facial expression) that something has been left out of the story, and many other cues. Picking up on these undercurrents, accurately, gives the coachee a sense that he or she is really being listened to. Good listening of this kind tends to open even more doors.

Questioning: Asking a good question in a coaching context often has a very different purpose from asking a question in normal conversation. Usually, we ask a question to get a piece of information that we didn't have before. In coaching we add questions that are intended to help the coachee think more deeply about an issue and to explore something that he or she had not yet considered. A normal question creates content information, whereas coaching questions create insight.

These are the basic building blocks of a coaching process and for creating a coaching culture. By adopting these attitudes and tools, you can really improve your chances for success in making change.

You might want to take this concept of coaching a little deeper and use it in a more structured way. We'll explore that in our next conversation.

But first, take a look back over these pages, and let your mind play with the following questions. We're not there physically to observe your responses or ask follow-up questions, so please be really honest with yourself.

- How much of this do you do already? That is, how often do you adopt a coaching attitude (regardless of whether you like to call it "coaching" or not) in the conversations you have with colleagues, members of your work team, your superiors, or a client?
- How often do you seek out this kind of support for yourself?
- How skilled are you at the kind of listening, and questioning, that we just described?
- What could you do to practice this more, and improve your skills? How could that be of benefit to you?
- How could you bring this kind of attitude and practice into a whole work team, to create a coaching culture?

12

A Way of Coaching: The GROW Model and Peer Consulting

KEY MESSAGES

- By intentionally structuring your coaching conversations, you can increase their effectiveness.
- The GROW model—Goal, Reality, Options, Will (or Wrap-up)—works well one-on-one or in groups and also fits well with the VISIS method of driving change.
- Peer Consulting is a group process in which teams reflect and learn together by helping an individual change agent generate new ideas and options.

We've talked about the basics of adopting a coaching attitude and creating a coaching culture in your change work. If you do this, even in just an informal way, you should notice a real improvement in how things unfold.

But sometimes it can be helpful to be more structured in your approach to coaching. Building on those basic principles and tools, a coaching conversation—or even a formal "coaching session," a conversation that you've scheduled, with the intention of approaching it in this way—tends to flow through a specific process in various stages.

There are several ways to describe this process, and they are often named with acronyms. Some use the CLEAR model: Contract, Listen, Explore, Action, Review. ("Contract" means the agreement you reach with each other about the purpose of the coaching session.) Others use a model called COACH: Concept, Outcomes, Application, Commitment, High Performance. Nobody seems to know exactly who first developed these methods, and as you can see, the flow is pretty similar in both cases.

We like to use a model called GROW—Goal, Reality, Options, Will (or Wrap-up)—as it fits very well with our VISIS model. GROW gives a general frame for structuring any coaching session, while VISIS provides tools to go a level deeper. Here's how the concepts match up:

Goal = Vision and Indicators
Reality = System
Options = Innovation
Will/Wrap-up = Strategy

Let's go through each of these steps in more detail.

Goal: Vision and Indicators

Every coaching conversation should start by coming to clarity about the **goal**. This is true for a single session, but it's also true for a whole coaching process (or series of sessions). Without such a goal in mind, the conversations easily lose focus and jump from one issue to another.

Defining the right goal is not always an easy task. People often have a tendency to feel bothered by immediate, problematic issues. This can lead to a superficial goal, at the level of solving the surface problem. For example: "I want to get rid of this problem I have of

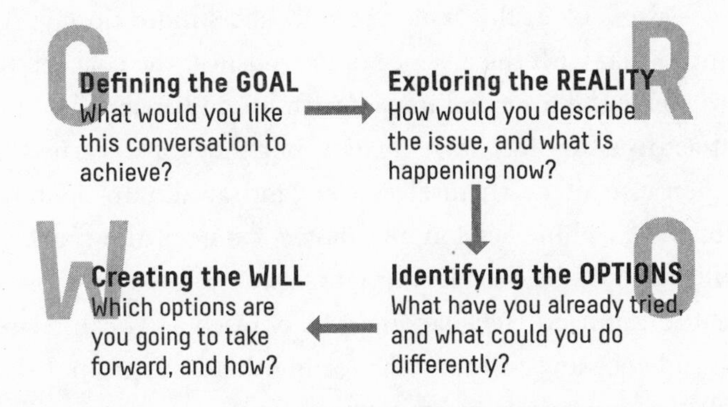

The Grow Model. This is our "standard" model for coaching (though there are no models that are truly standard). We like it because it aligns well with the VISIS Method. When you do coaching, be sure to use a model that both you and the coachee are comfortable with.[13]

feeling nervous when I give a presentation." Getting rid of a problem may be a reasonable thing to want to do, but it's hard, or even impossible, to find your way to a new set of options by imagining the *absence* of something. The coachee, as well as the coach, should have the mutual clarity to describe the goal to which the conversation aspires in a more substantive and positive way. This is something like having a common vision for the conversation: what they hope the coaching session will achieve.

Attempting to describe that goal or vision more effectively brings us to the precision of **indicators.** How will we know that something has changed? What signs should we look for, to know whether the coaching session—and the follow-up action that results from it—is successful?

To continue with the example of how to help someone give presentations that have more impact (which is often a critical skill in the context of change processes), the first attempt to define the goal might be that the coachee says, "I don't want to feel nervous giving the presentation." To define the goal more clearly in terms of vision, the coach might ask, "How *do* you want to feel?" And then: "What does a successful and high-impact presentation look like for you?"

Let's also look at this issue of nervousness more closely: Are we convinced that with the absence of nervousness the coachee will be able to create the intended impact with his audience? Even professional actors often feel stage anxiety, but they have learned to use this energy to make themselves alert and awake. In defining the goal of the coaching session, the *intended state* of the coachee as a presenter should be described; that is, the state that is most likely to create an impact. The goal might be phrased as, "I can relate well to the audience and maintain the feeling that I am grounded in my knowledge, experience, and self-esteem."

Probably you can also sense that by reaching that state the likelihood of convincing an audience with your presentation will rise. In terms of the API model we introduced earlier, the coachee has described a state of stronger *Presence* ("I can relate well to the audience") that is also rooted in *Authority* ("grounded in my knowledge, experience, and self-esteem").

To make this description of the goal even more precise, we need to identify some indicators that will help the coachee know that she or he has achieved the stage. Indicators for relating well might be to keep eye contact with people while presenting, talking directly to people by focusing on several individuals, and keeping a friendly attitude toward the audience by reminding oneself that the audience also wants the presenter to do a good job.

Indicators for feeling grounded in one's knowledge, experience, and self-esteem could include having an upright body position, breathing into the belly (rather than shallow breaths into the upper chest), and keeping in mind that many people see you as an expert on this topic, and the audience's knowledge about it is very limited.

Reality: The System

In this phase the focus of the coaching conversation shifts. The coach wants to help the coachee understand the situation more deeply. We use the following little frame to look for that deeper understanding.

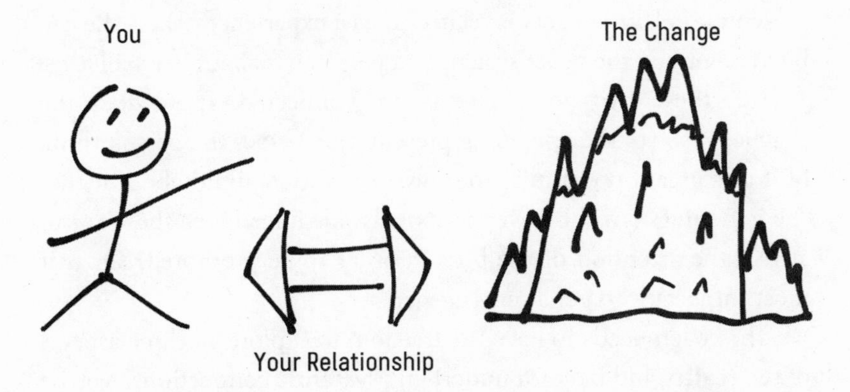

The three elements of any situation faced by a person being coached (the "coachee"). Two of them are obvious: *You*, meaning the person making the change, with all her skills and experience; and the specific *Change* that that person is trying to make (think of the change as "the mountain she is trying to climb"). The third element is more subtle: it's *that person's Relationship to the mountain.* How she reacts to it; what attitude she has when trying to climb it. It is very helpful to keep these three elements as distinct, exploring them separately from each other, as well as exploring the ways they interact.

Every problem somebody is facing can be divided into three different elements:

- The *Change* is the objective situation, the challenge out there in the world. (That's the mountain. Sometimes we call this the "Issue.")
- "You" means the *person* and refers to the basic feelings, attitudes, skills, and other aspects of the coachee.
- The *Relationship* is how that person *actively relates* to the Change or the Issue: what thoughts she thinks, what actions she takes.

In our example, the Issue is the actual presentation. This includes a specific audience with a certain composition or status, the content, maybe some presentation slides, the room and setting, and so on. Looking to the person, we see the feeling of anxiety about the presentation. This anxiety is related to past experiences, skills she can build on or not, the belief system that person has about her skill level.

But these two outer aspects are connected: How does the coachee *relate* to the upcoming presentation? Does the person think of it as a great opportunity to grow, or is she so afraid about it that she is hoping it will be over as soon as possible? Does she want to attract the attention of the boss in order to be promoted? Or is it most important to get a message across?

The coach needs to have the freedom to explore all three aspects of the **reality** and look for underlying **systemic** connections. Maybe low self-esteem leads to trying to get the presentation finished as soon as possible, by only having a few slides. In that case we could intervene with suggestions and options (see next section) on all three levels, since they are all connected.

For example, a longer and more detailed presentation about the Change or Issue might help the coachee root herself in her Authority, which then might increase her positive thinking about holding the presentation (her Relationship to the Change). But it's not usually that simple: a longer and more detailed content approach (Issue-focused) might increase the anxiety level (for the person), which might require some additional focus on relaxation techniques, such as deeper breathing during the presentation (the Relationship). So these aspects are not simple and one-dimensional: the cause and effect patterns create complex, **systemic** interconnections. This part of the coaching session usually takes a bit of time to really explore and understand these connections.

Options: Innovation

Coaching involves adopting an organic, systemic, and exploratory paradigm, or way of thinking, not a linear or mechanical one. We're not trying to fix someone, or to replace one behavior, feeling, or belief system with another. We start with appreciating what the coachee already has, and acknowledging that there are very good reasons for every behavior, feeling, or belief system. They all make perfect sense when put into the frame of the coachee's personal history.

The goal of coaching is to build on what's already there and to help the coachee develop new **options** within that existing behavior, feeling, or belief system. These options will, in the long run, give the coachee more freedom to choose among different ways of reacting or relating to a situation. The old habits will always be one option, but as time moves on, maybe the old options are no longer appropriate. Here's an example.

> *Axel: At school I was a bad student, and I always tried to hide and not be put in the limelight. So when the teacher asked an open question to the class, I automatically tried to hide behind the backs of my fellow students.*
>
> *So I still have the habit of holding back and giving space to others, which is also a good quality. But as a trainer and consultant, I often need to stand up and present. So I really needed to develop alternatives and new options for my behavior.*

A good option is seldom more or less of the existing behavior. Nor is it often the *opposite* of the existing behavior. It is rather something new and **innovative**. So when, in a coaching process, you shift from Reality to Options, you also need to shift your own mind-set and attitude. In exploring Reality, you need to create an atmosphere of precision, understanding the "unsaid" and seeing the connections. In this phase the coach is mainly listening and driving understanding through open questions.

Exploring options is a much more creative and mutual brainstorming process, which involves generating ideas and even allowing a certain kind of wildness being present ("Here's a crazy idea . . ."). As long as the coach is not attached to his or her ideas, the coach can be as participatory in this brainstorming as the coachee. The right idea will pop up, and it doesn't matter who formulated it in the first place. It only matters whether the idea helps the coachee to feel, think, or act differently in a given situation.

Whether an idea fits or not is easily detected by the reaction of the coachee. He should feel some kind of excitement or relief when he thinks about the idea. If the idea doesn't touch the coachee in some way, it's most likely not a good option.

Axel: I would like to connect this to a real coaching situation I had that involved helping someone overcome his fear of presenting to audiences. We had developed some minor options to help the coachee stay connected to his self-esteem. One option was for him to ask for feedback from the members of the change team about the relevance of what he has to offer, based on his expertise. He wanted to write all the feedback on a piece of paper and keep it in his inside jacket pocket, at the level of his heart. Whenever he felt insecure, he could think about that feedback and connect to the feeling that he had good colleagues who believed in his Authority.

But then, for staying present with the audience, he developed a breathtaking new option. He decided not to do a slide presentation at all, as he had planned, but to create an open dialogue about the issues with the participants. He would summarize the main facts as a one pager, which he would hand out to everybody at the beginning.

This option was a true innovation *for him. Of course, it needed to be planned later in more detail, but the breakthrough idea came during the focused coaching conversation.*

Will: Strategy

"Will" refers to the coachee's intention or commitment, after the coaching session: what he will do, and when. On one level this phase tries to make sure that the option or options that have been identified will be put into practice. But the pure intention to do something is sometimes not enough to ensure that the new and innovative options move to become used behaviors, feelings, and thoughts. So **will** alone, without a **strategy** to build those new options into a routine, is not enough.

Axel: We made a little plan with the coachee I described above, before he arrived at his innovation, to really focus on "what and when." It included a time frame for getting feedback from his colleagues and then writing the essence of it into an "appreciative letter" to himself. After he had that, we anchored it to a routine in a different coaching session. When he started to feel his self-doubts rising, he thought about connecting himself to that letter, inwardly.

Another "what and when" was condensing his slide contents into a one-pager. He needed to put some time into that and discuss it with a close colleague, who also knew the content.

To help him realize his goal of creating a dialogue with the audience, we rehearsed his presentation in a coaching session, during which I played the role of the audience and he tried to engage me on that topic. Then I could give him feedback on his API—Authority, Presence and Impact—and give him some on-the-spot tips.

For improving performance, rehearsals are nearly always a good idea. As change agents we often just rush into situations without actually practicing first. When you have the time, you can do one or two rehearsal sessions with your team before you run an important meeting or give an important presentation. Imagine how often actors rehearse a play before they go live! We as change agents and change teams could learn a lot from their professionalism.

The Peer Consulting Process

Coaching is, of course, a kind of consulting, a word that simply means "thinking deeply together." Sometimes you might face a difficult situation in a change process, more at the level of a whole team, where you could use a consultant—but don't have the time or resources to engage one. This is where a coaching culture comes in. You can face these difficult situations together.

The process is called Peer Consulting. It is an intervention tool to help a group of colleagues reflect, learn, and use their mutual intelligence and experience to help an individual create new ideas about how to act in a specific situation.

This is a well-tested approach that takes about one hour of team time, in a very structured format. It works best with a group of approximately six to eight people. You need to have somebody, the presenter, who has an open issue to discuss among the group of colleagues and one person who is facilitating. The rest of the group will be listening and discussing.

To begin with, the presenter takes some time to describe the issue and make some notes about it on a flip chart or whiteboard. The issue needs to be:

- *Real and current.* Don't present something hypothetical, or something that you faced a year ago. Although a group might not know this, the fact that it's not real or current will be felt intuitively (the presenter will not have the same "energy" in presenting it). In this way groups are much smarter than we often think and can't easily be fooled. You and the group will get the most out of a Peer Consulting session if together you have the opportunity to create a real solution, one that can have a real impact.
- *An issue of personal influence.* There are many different issues that bother us in a change process. You can divide them into three categories:
  - Issues on which we can have no personal influence
  - Issues that we might be able to influence indirectly
  - Issues that are under our direct control and influence, which can change if *we* change them

Having this in mind, the notes that are presented by the presenter could be structured into four areas.

Headline: Summarize the whole issue in one eye-catching line, the way tabloid newspapers would do. The headline should be to the point, and provocative.

People involved: In the next section of the flip chart, focus on the people who directly or indirectly play a part. You could note information about their

- Position and function
- Age and gender
- Relationship to the issue

Description of the situation: Here you focus on the scenario that bothers you. Please be as precise and current as you can and describe real incidents, which involve such questions as these:

- What happened when?
- What was said and done by whom?
- What form did these events take? That is, were they in a private conversation, a larger meeting, or some other situation?

Personal Request: To help the group focus and come up with good options and solutions, it is important that you have a real request, formulated as a question. To come up with your request to the Peer Consulting group, think about these two questions:

- What is my open question?
- What do I want to achieve with this session?

Structure of the Peer Consulting Process

As we mentioned earlier, the benefits of this process are produced partly through having a very clear structure, in terms of both focus and time. So the facilitator should really play this role very attentively and actively. In this process there are six specific phases:

1. Presentation—10 minutes
2. Questions—5 minutes
3. Analysis—25 minutes
4. Review—5 minutes
5. Options—10 minutes
6. Feedback and outlook—5 minutes

1. Presentation of the case. The presenter gives all the relevant information (as described earlier) and describes the request to the group. The presenter should make sure to:

- Provide the information in as concise and precise a way as possible.
- Be concrete and specific, and distinguish between observations and interpretations.

The group of colleagues just listens during this phase. Besides focusing on the content, colleagues also pay attention to the *way* the presenter talks about the case.

2. Questions—Collecting further information. In phase 2 the colleagues ask clarifying questions.

- Ask questions about the people involved and about what actually happened.
- No rhetorical questions allowed! The colleagues are just seeking information and clarification.
- The presenter stays on topic and answers as briefly as possible.
- The facilitator needs to be alert and intervenes when necessary to make sure that the group only asks clarifying questions and does not start to discuss the issue or already start looking for solutions.

3. Analysis. Important: In this phase, the presenter sits *outside* the circle of the colleagues talking about his case and just listens and takes notes. With all the given information in mind, the colleagues analyze the described situation from different perspectives.

- The presenter listens to what is said and notices any reactions that come up inside himself.
- Colleagues analyze the case without paying attention to the presenter.
- The case should be analyzed from different perspectives. For this the model of "Person–Relationship–Issue" that we introduced earlier could be very helpful.
- Different points of view should be considered at the same time. It is not necessary to reach a consensus or a conclusion that is shared by all participants.

4. Review. After the analysis time is complete, the presenter briefly shares his or her impressions or reactions with the colleagues and corrects any possible misunderstandings.

5. Options. Important: Just as in the Analysis phase, the presenter sits outside the circle of the colleagues, listening and taking notes.

- The colleagues brainstorm and collect different options for the case.
- The presenter listens to what is said and notices any reactions coming up inside.
- Different possible options or solutions are given names or titles by the group.
- Crazy ideas are very welcome.
- The group members don't discuss or criticize the ideas that come up in the brainstorming. They just keep generating other ideas.

6. Feedback and outlook. Finally, the presenter gives feedback to the group:

- What have I learned?
- Where did I react strongly—agreeing or disagreeing?
- Which ideas will I try out?
- When will I do it?

Afterward, the whole group can take a moment to reflect on what they have learned. Our experience has shown us that while the experience of one individual is sometimes just personal, it can often be an indication of a systemic issue, which that individual was simply the first to feel. In such cases, every colleague will have something to learn from this situation.

The Peer Consulting Process can help the whole change team to reflect continuously on systemic patterns within the change process and sharpen their individual strategies to deal with these patterns.

* * * * *

Now you have a few new tools in your toolbox. Structured coaching conversations and Peer Consulting processes can be very powerful ways to advance change, because they help unlock more of the creativity and insight that we all have.

Take a minute to think about this to yourself. How might you put tools like these to work? How could coaching, and creating a coaching culture, help you?

If you feel that doing something structured like this might be uncomfortable at first, can you think of some way to reduce that discomfort? Can you practice with a few friends, for example?

If you feel that coaching of this type is something you would strongly benefit from (acting either as coach or as coachee), do you have any resources around you to deepen your engagement with these ideas? Do you know a good professional coach, for example?

What are your next steps? How can you put these ideas into practice?

13

Approaching Sustainability Transformation

KEY MESSAGES

- The challenges of change in a sustainability context are usually larger scale and deeply structural. This makes them more difficult—but they are still very possible.
- The skills of communication and coaching others become even more critical to success. Sustainability transformation is a "team sport."
- By practicing patience and building your storehouse of experience, you can increase your ability to know when to actively intervene and when to just "let things happen."

Until now we have not focused much on the specific topic of sustainability. That's because we believe the challenges faced by change agents are very similar across most change topics. Initiating any significant cultural shift has a lot in common with the dynamics of change for sustainability, and most sustainability transformations are cultural transformations at the same time.

Now let's take a specific look at sustainability from a change perspective.

A *sustainability change process* usually involves a high degree of complexity, because it is trying to improve several kinds of performance at the same time: environmental, social, and economic. Even if the process is focused on a specific sustainability topic—say, reducing carbon emissions, which most people see as an environmental issue—it nearly always has many social and economic dimensions. Usually, it is not possible to just flip a switch and reduce emissions. You usually have to change a number of system elements, involving everything from energy technologies to accounting systems to people's habits.

When the change you are trying to make goes even further and requires a comprehensive shift in behavior, perception, and attitude—a *transformation*—the process becomes more challenging. And many, if not most, sustainability change processes fall into this category.

The exceptions are sustainability change processes that are building on transformations that have already happened (for example, incremental improvements made after a big reorganization), and programs that are just trying to polish up an organization's reputation (trying to make the company just a little cleaner and greener, for example).

Why have we been urging you, in this little series of conversations, to think so much about people, patterns of behavior, and culture? Because when you are involved in a transformation, "Culture eats strategy for breakfast."

This quote comes from the famous organizational consultant Peter Drucker. It sums up the situation very well: If you have an important change to make in an established culture, you need to be very, very smart about how you implement it. Otherwise the people and the systemic patterns in that culture are likely to resist that change very strongly. Stable cultural systems have an extraordinary ability to maintain their current status or to go back to their original status after being disturbed. In simple terms, when you push them they try to "snap back" to where they were before.

Meanwhile, some of the core beliefs behind sustainability—such as systems thinking, ecological limits, and social responsibility—are often in conflict with existing practices in organizations, or even in society as a whole. Take, for example, the idea of long-term planning perspectives. Most business organizations, indeed most national economies, are driven by the quest for short-term results. Making changes just in that one area, shifting people to longer-term planning horizons, can be very challenging. Why? Because trying to change that "one thing" actually involves changing a lot of *other* things about how the organization works—as well as changing how many of the people affected by that organization think and act. Many different forces of resistance come into play.

But transformative change for sustainability is certainly possible. Otherwise, why would we be here, talking with you in our virtual café?

Here is an inspiring example of transformative change in a large organization that shifted toward long-term thinking and overcame that resistance. As is often the case, it has to do with good leadership.

Alan: In 2014 I had the honor of interviewing Paul Polman, the CEO of Unilever, live on stage when he was given the Gothenburg Award, Sweden's "Nobel Prize" for sustainability. One of the many transformative decisions he is famous for was shifting his global company away from quarterly financial reports to only publishing annual ones.

This sounds like a relatively tiny change, but in the corporate world—where investors and stock prices and all the rest of it are driven by quarterly financial reports—it was huge. It was also necessary, Polman realized, for making any of the other big changes he wanted to pursue to turn his company toward sustainability. That's because changes that result in long-term sustainability often look more expensive in the short term than business as usual—even though they lead to long-term health for a company. Companies often shy away from sustainability investments to keep their quarterly earnings looking good.

Immediately after Polman's announcement, 10 percent of Unilever's stock was sold off by worried investors. That's okay, said Polman. We only want investors who have a longer-term perspective.

And indeed, new investors bought shares, and they did well as a result. Not only did Unilever attract global attention for its comprehensive Sustainable Living Plan, which transformed the company in other ways, but also its stock price performed well in the following years. Polman's brave decision about financial reporting (and many other decisions he's taken to change his company) proves that transformation is possible.

Practicing sustainability involves expanding our focus to consider the long term and the whole system. But our current culture, in most organizations, in most of the world, pushes us to do just the opposite. We tend to focus almost exclusively on the success of individual people, departments, companies, and so forth when we also need to think about the success of the whole *context* on which these things depend: nature and its resources; people and their well-being; the social and economic systems we have built up around us over generations.

Practicing sustainability means learning to think outside the box of our immediate concerns. Making a profit or meeting a target in the next quarter year is fine. But how do we make sure that we can keep doing that, over years and decades? How do we make sure that success today, for us, is not actually leading us to serious problems tomorrow, for us and for many other people?

Obviously, once you start asking questions like this, many issues come up, and many of these issues have to do with environmental and social concerns. These are the classic issues that sustainability change agents are usually working with. This list of issues includes things like global warming and the carbon dioxide and other greenhouse gas emissions that cause it. It includes other environmental problems, caused by organizations (and people) that believe they have to take resources out of nature, or dump wastes back into nature, or even destroy nature, just to survive. And it includes a broad range of social concerns, including issues that seem very far away from us in both space and time, such as how people on the other side of the planet are doing, and whether their children are growing up with a chance for a better quality of life.

In a business or organization that is focused on short-term, immediate success—and that is also embedded in systems that keep pushing it to think this way, supported by the well-established habits of millions of people—facilitating change for sustainability is certainly not easy. And successfully bringing about real transformation can look almost impossible.

And yet there are many stories we could tell that are similar to the one about Paul Polman and Unilever. *Sustainability transformation happens.*

But how does it happen?

The answer is deceptively simple, and it builds on everything we have been talking about so far. It's all about having a clear vision, a good idea (or set of ideas) to promote, a solid strategy for implementing them, a lot of courage and flexibility . . . and the willingness to look inward, listen deeply, and spend time listening to others. It's about getting to know yourself as a change agent and getting to know the people you are working with as well as you possibly can, then working *together*. Sustainability is a team sport.

This is neither the time nor the place to get into the science of sustainability. Nor do we want to tell a lot of stories now about companies, cities, governments, NGOs, and other organizations that have successfully implemented a sustainability change or larger-scale transformation. There are many other books and websites full of such stories, and one of us (Alan) has written a few books about these topics.[14]

Instead, we would like to use this part of our conversation to dig a little further into the *idea* of sustainability transformation and to think with you about what's required of us as change agents when we are trying to make transformation happen.

Transformation = Changing System Structure

When an organizational system is unsustainable, it means that if it keeps doing what it normally does, it will eventually have big problems.

Consider a company that is very dependent on fossil fuel energy (there are many of these, including car companies, airlines, and the financial companies that serve them). The company might look quite healthy and strong today. But if you adopt a long-term perspective, and a broader field of concern than the company's own boundaries, the picture changes a lot. The world as a whole has woken up to the inescapable reality of climate change. Governments are slowly changing their policies and incentives to steer more and more toward renewable energy. Investors and big banking institutions are starting to worry that most of the oil and other fossil fuel will have to be left in the ground, placing doubt around the future value of those economic assets.

From a sustainability perspective, a company that is very dependent on fossil fuel energy will not be able to keep doing what it's doing in the longer term. Small changes will not be enough, either: just getting a little more efficient, or creating a greener profile, or becoming a little more engaged with community issues will not help. They are nice things to do, and they might even be important to do. But they are still just tinkering in relationship to the real issues.

Let's look at another energy example. Earlier in the book we mentioned the *Energiewende*, which began to be implemented in Germany (where Axel lives) as a reaction to the nuclear catastrophe in Fukushima. To some extent this *Energiewende* seemed very sudden, and it was even initiated by a conservative government that was not known for promoting green ideas. But actually, there was a movement toward renewable energy and green thinking in German society that had been going on for many, many years already—to such a degree that the Green Party had sometimes been leading the country in a coalition with the Social Democratic Party.

Even though the issues of energy and climate were already on the table in Germany, the decision to implement *Energiewende*—the nationwide transition to energy efficiency and renewables—hit the biggest energy companies totally by surprise. Nuclear power stations were their cash cows, and they didn't have a plan B in their desks for this sudden turn of events. These companies were among the biggest in Germany, and suddenly they were in very big financial difficulty—because their cash cow was unexpectedly slaughtered.

There is another layer of public catastrophe connected to this: a likelihood that these companies have not put aside enough funds for rebuilding the existing nuclear power stations or handling the nuclear waste—and definitely not enough for the tens of thousands of years that this extremely dangerous, radioactive, toxic waste will be around, demanding very special attention and high levels of technical skill and investment.

Milking the old cash cow seemed to be fine, but taking care of the side effects and long-term impacts was not part of the plan. This is just one example of a deeply unsustainable pattern, but it is not untypical of our times and our current economic mind-set.

To be sustainable, energy companies will have to start reconsidering the whole *system structure*: what kinds of products and services they sell, how they sell them, all the economics around that, and all the communications to people about what the company is doing or intends to do—not to mention the mind-sets (mental habits) and skill sets of its employees.

Here's a small example.

Alan: One client we've worked with is a large company that sells, among other things, cars. (I talked about them previously.) But they were slow to adopt the new technologies that are making cars greener, such as hybrid engines that combine battery power with traditional petroleum-based combustion engines.

Of course, hybrids are well established in many parts of the world. What was the obstacle in this particular case? It turned out to be a mind-set. Most of the people working at the company believed the hybrid cars were somehow inferior to "normal" cars when it came to performance. As this was also the prevailing mind-set in their market, it was hard to start changing customer buying habits if the people selling the cars had this kind of mind-set, too.

As it happened, the new hybrids were actually better (faster in acceleration, for instance) than the "normal" cars, but nobody believed that yet. Then the sustainability director came up with a great idea: he used his own budget to upgrade the cars that all the senior managers drove to work. They already had pretty nice "normal" cars, but he made sure they all got nicer luxury models—which were also hybrids.

This changed the mind-set: suddenly hybrids were an "upgrade" that everyone in the company used and appreciated. This made them easier to sell as well.

Remember the process we described around coaching, in the last two chapters? Maybe it's easier now to see why we think there's such a strong connection here. The idea in the above story emerged out of a coaching relationship with that client that involved many sessions of thinking together about all the many aspects that were standing in the way of progress and looking for creative new options.

Sustainability transformation usually involves changing *many* aspects of a system, in terms of physical processes, social routines, and mental habits. It brings up many challenges that can only be solved if we adopt a very reflective, open, learning attitude—individually, but also in our work groups, change teams, leadership roles, and other professional relationships.

The first thing that sustainability transformation requires of us is the recognition of the scale of the change: big; comprehensive; containing within it many smaller changes, opportunities, and problems to solve. Facilitating a sustainability transformation means facilitating *many* moments of change—and supporting others to do the same.

Hurrying Very Patiently

The second thing sustainability transformation requires from us as change agents involves getting comfortable with a certain kind of paradox. On the one hand, the situation is urgent (or could quickly become urgent), so changes need to happen as quickly as possible. On the other hand, change of this kind is so big, and so complex, that it usually can't happen very fast—unless a catastrophe forces it to happen. And even when change is forced, most sustainability transformations at the level of large organizations or communities still play out over several years. When talking about large cities and whole countries, the scale is usually a decade or more. Germany's *Energiewende*, for example, has gone from concept to wide-scale practice over a period of more than twenty years—which is nonetheless considered very fast indeed.

As change agent you will have to be *very* patient. Facilitating transformational change is often like watching a glacier melt or a tree grow: there are times when it seems that nothing is happening, but when you take a photograph a few years apart, you see enormous changes. You need to keep reminding yourself of this reality, when things seem slow. And don't forget to look backward once in a while, to see the progress that you've made over the scale of months or years.

But also, during any process of transformation there will also be many moments of fast action, and even crisis, when things are truly happening quickly, and you have to hurry to keep up. Those moments require another kind of patience, which people sometimes call "centeredness": a calm resilience that allows you to maintain your reflective attitude and not lose sight of the big picture.

Cultivating patience is one of the greatest gifts you can give yourself as a change agent, because you will need plenty of it.

The Continuous Identification of Leverage Points

Do you remember our conversation about systems and leverage points in chapter 8? Transformations are about changing systems, and that means knowing *where* in the system to focus your efforts, or to help other people focus their efforts. Systems that are stable have a strong tendency to try to return to the way things used to be. It's not often that you can just set a transformation in motion and then step back and watch it unfold. Instead you or someone else, or usually a group of someones, will have to continuously engage with that system, helping it overcome small hurdles and solve sequences of challenges on its way to a new, stable state.

That means that you and your colleagues will be engaged in a nearly continuous search for leverage points. And as usual, these leverage points will take many forms. Sometimes you will need to look at new technologies or practical solutions. Sometimes you will need to look at where communication is or is not happening. And sometimes you will have to focus your attention on the belief systems of the people involved.

Here's an example.

> *Alan: Many years ago we were engaged by a large foundation to give them strategic advice on their environmental funding strategies. My team and I worked with them over several years as they rethought what they were doing, and we helped them find better leverage points for change in the surrounding community. They*

were having good success, too: by shifting from a "save nature" communications strategy to an "improve human health" one, they had dramatically increased local engagement in cleaning up waterways and taking other pro-environment actions.

But they had neglected to change their own systems internally. They asked us to help, starting first with a plan to make their kitchen as green as possible. We researched this thoroughly and came up with a comprehensive approach that looked great on paper. It even involved beautiful handmade plates, fashioned from recycled glass, to replace their habit of using paper plates.

But when we came back six months later, nothing had happened; our plan was not being implemented. Why? Nobody knew! We had to search carefully to find the leverage point and unlock the change. There was no problem with the technologies proposed, the cost, or the other obvious obstacles.

It turned out that the place where we needed to intervene was in one person's belief system.

The president's secretary believed that "environmentally friendly" dishwashing machines would not get the dishes clean enough. This person was quite afraid of germs and of getting sick. So she had simply prevented the plan from going forward by quietly and consistently moving the document to the bottom of the president's in-box.

Once this leverage point was identified, our clients sat down and listened carefully to her concerns, then showed her the independent testing data on the dishwasher we had proposed: it was hospital quality, and it sanitized the dishes better than any household machine. This satisfied her and unlocked the process of change.

At every moment in a sustainability transformation process, you need to be prepared to reanalyze the system—in all of its aspects!—

and figure out what's blocking change, or where change can be accelerated or facilitated.

Actively Let It Happen

Here is another paradox: while we just described the need to be continuously on your toes and ready to intervene in multiple ways over long periods of time, there are times when you need to *not* do that.

This is not about being passive. It is about being observant. When you notice that things are happening under their own power, that the dynamics are shifting in ways that may be turbulent but favorable, you may need to hold yourself back and control any tendency to step in and try to facilitate, manage, or otherwise encourage things to happen faster (or happen differently). Systems may often be resistant to change and ready to snap back into their previous ways of being. But they also have a lot of their own wisdom. When change becomes inevitable, they often *know* how to change.

Some readers may be aware of the Chinese book of wisdom, the *Tao Te Ching*, by Lao-tzu, which is about twenty-five hundred years old. It introduced the world to the concept of *Wuwei*, which is often translated as "doing by not doing." Being an agent for transformation often requires the ability to let something happen, without being visibly active.

When should you jump in and try to help move things along, and when should you actively hold yourself back, watch, and keep letting whatever is happening run its course? How do you know?

This type of discernment requires a lot of that attentiveness, reflection, and patience we've been talking about. As a change agent, whether you are working internally as part of the organizational culture or externally as a consultant or advisor, you simply have to sharpen your observation skills.

And here is the hard truth: there is no real substitute for trial and error. You have to have the confidence to try. Sometimes you will make the wrong decision, and usually that error becomes obvious in retrospect. You'll learn.

The important thing is, whether you choose to intervene or to stand back and let things happen, to make that choice *actively*, consciously, so you know *why* you made that choice. Then when you observe the results of the choice, you will be better able to sort out whether it was a good one.

Through experiences like these—actively intervening or actively stepping back and letting it happen—and through studying how change happens in other contexts, you can build up a storehouse of knowledge that will help you get better at making the right decision. You will develop a feeling about these things, what we usually call intuition, which is actually the result of your brain processing a tremendous amount of information that you are probably not even conscious about. By staying actively attentive to what's happening around you, and paying an equal amount of attention to your internal signals, you will start to "just know" what the right thing to do is, in any given situation. You won't always be right, but with time you will be right more often.

* * * * *

These were four important things that we've learned over the years about how to approach the process of sustainability transformation: with respect for the size and complexity of change, with great patience (even when you are in great haste), with a readiness to search continuously for leverage points, and with an equal readiness to hold yourself back from intervening. This is important when things are already moving forward in a good way.

But it is even more important at certain specific moments when the process may seem to be stuck. Sometimes you just have to remain still and grounded in trusting the intelligence of the systems you are working with, embracing a kind of humbleness. As a change agent you and your knowledge and expertise are very important, but you are only one source of making a difference. In any system there are always more sources, and many of them are difficult to see. You just have to remain patient and let them do their work.

At least, that is what we have experienced in our years of working with transformative change processes.

What about you?

What tips or advice would *you* give to someone who was trying to make a transformative change in some kind of organizational system for the good of the whole?

14

Parachuting Cats into Borneo and Other Cautions

KEY MESSAGES

- When you are trying to change complex systems, it is good to approach them with humility—and with the expectation that something is likely to go wrong.
- The change agent has to be constantly on her or his toes, ready to adapt, learn, and try something new.
- The story of the "Parachuting Cats" can help you cultivate a sense of healthy caution—but don't let caution stop you from taking action.

So far, we have really focused on encouraging you. And we still want to encourage you! Making change is often tough work; change agents will often meet with obstacles and doubt, both outside in the world and inside themselves.

But there are a few cautions we would like to share with you. As you probably know, not every change turns out to be a good idea. Some are dead ends, paths that end up leading nowhere. Others just do not work as you expected them to, sending you back to square one to try something else. And some changes, while they seem to solve the immediate problem, actually make things worse.

Here is an old story that we often tell to illustrate this.

In the early 1950s, on the island of Borneo, the relatively new World Health Organization took on the challenge of eliminating malaria. They sprayed the pesticide DDT to control the mosquitoes, which transmit the malaria parasite. Malaria rates declined. That was certainly the effect that the change agents were looking for. But unfortunately, the DDT had some unexpected side effects, which took them very much by surprise. Those side effects illustrate the benefits of systems thinking.

First, the DDT also killed a certain wasp, which usually laid eggs in a certain caterpillar. Without the wasps around, the caterpillars had no natural predator, and their population exploded. One of the things they liked to eat was the thatch on the roofs of the village houses in this region. Roofs began falling in.

The WHO rushed in to fix that problem by helping the villages replace thatch roofs with tin ones. But then came another unexpected problem: the DDT had killed all the cats.

Suddenly, with no cats around, the rat population exploded—and the rats carried with them diseases such as typhoid and one strain of plague. Now the WHO was facing outbreaks of new diseases—caused by its original initiative to stop malaria.

Here the story gets a bit surreal but is apparently true: to take care of the rat problem, the WHO gathered cats from other areas and parachuted them into remote villages with the help of the British Royal Air Force.[15]

The point of the story of the parachuting cats is that we always have to be a little humble about the change processes we are introducing. Most organizations, communities, and planets are pretty complicated. We usually cannot predict the full impact of our change efforts. There are bound to be some side effects.

What can we do to reduce the chances that we will be forced to parachute cats into Borneo in our own situation?

Here's where the concept of a coaching culture comes in again. There are also many other phrases that bring up other aspects of this same reflective attitude, such as "learning organization." Basically, we always have to be ready to ask questions, listen deeply, search for new understanding—and try to let go of the feeling that we must know exactly what's happening, or going to happen. This is just not possible, of course.

And yet, in organizational life especially, we often pretend that it *is* possible. We make strategies and plans with key performance indicators and often with very little provision for uncertainty, ambiguity, and complexity.

In fact, there is an acronym (originally from the military) that has recently come into wider use in business: VUCA, which stands for Volatile, Uncertain, Complex, and Ambiguous. (You can read more about this in the *Harvard Business Review*.[16])

As a change agent, you should bear in mind that most situations you face will have a lot of VUCA characteristics. *Volatile* means that the situation can change quickly and unexpectedly. *Uncertain* means you can't predict outcomes reliably. *Complex* means that often you won't even be able to understand what's going on, because there are so many things in motion. And *Ambiguous* means that there will be times when you won't be able to tell whether what's happening is good or bad or just different.

In short, you really have to be on your toes, all the time, ready to learn, adapt, and try something new. (Or occasionally, as we wrote in a previous chapter, have the discipline to hold back from trying anything new and let events unfold.)

Now, we could continue here and make a long list of things that can potentially go wrong with your change process. But as we said, our chief aim is to encourage you, not to make you worried.

Nearly all the skills, tools, attitudes, and approaches we have described in this book, our "virtual café," are the same ones that will help you deal with the challenges you face as a change agent—both the small challenges and the large VUCA-size ones.

Just keep this one image in mind: parachuting cats.

Let the parachuting cats be a symbol for how we need to be respectful of the complexity of the systems we're working with, trying things out, listening, learning, and thinking; asking other people to help; being ready to test our assumptions and belief systems and adjust them—and our resulting action strategies—when the information changes.

Which it will—often.

How about you? Do you have any parachuting-cat stories of your own? Times when you tried to make a change (or observed someone else doing so), only to find yourself dealing with unexpected side effects, or even things spinning out of control?

How did you respond? What did you learn from that experience?

15

The New Leadership for Change

KEY MESSAGES

- Corporate governance provides us with many lessons for what leadership means, and requires, in a sustainability context.
- Unexpectedly, looking at corporate leadership leads us straight into a confrontation with one of the ultimate mysteries in human life: consciousness.
- New theories of leadership make it clear that successful leadership for change and sustainability means cultivating the evolution of consciousness—our own, as well as that of those around us.

We hope that by now you have gained some new understanding about the complexity that a change agent must deal with. She or he has to handle a multilayered process, involving the interests of different stakeholders and their potential conflicts, while supporting his peers to create impact, even when systems have the tendency to resist change or return to their original shapes.

Dealing with this complexity is already a huge job. How much bigger does it become when your objective is to create a more sustainable world, in a culture that has the tendency to sacrifice the long-term well-being of the whole to satisfy the short-term desires of individuals? Individuals in this context means not only individual people but also individual units, such as companies or countries, putting their own interests ahead of the needs of all other units.

Let's consider individual companies. They are inherently competitive, yet any mainstream economist will tell you that companies actually need *other* companies to also succeed in their market—otherwise there is no market, no pressure to improve and become more efficient. Consequently, even at the level of ordinary competitive thinking, total selfishness does not work: we need our competitors if we are to excel. This is why you so often see a spirit of comradeship in elite sports, with even opposing players really taking care of each other and appreciating each other. They know, deep down, that they need each other.

We think that everyone knows, deep down, that we all need each other, and that we all need to take care of our planet as well. But reminding people of that, and successfully motivating them to change their behavior—and change their systems—so we really are taking care of these things requires leadership.

For most of this little book, we've been focusing on being a change agent. Leadership and change agentry are not the same thing. You don't always have to be an identified leader to be an effective change agent. In fact, some of the most successful change agents remain rather invisible in the change process, working behind the scenes to make sure communication is flowing and ideas are moving where they need to go.

And leaders are not necessarily change agents. In fact, many great leaders made their mark by helping people hold steady and resist some change that was being forced on them.

But often, in sustainability work especially, change agentry and leadership do go hand in hand. Change agents must develop their skills of leadership to persuade people and motivate them forward, and leaders must act as change agents, bringing new ideas and practices into their organizations.

Let's look at two sustainability-related ideas that are examples of situations in which these two skills must be combined if the idea is to succeed.

Changing Corporate Governance

We've already mentioned the story of Paul Polman at Unilever and his decision to change that company's planning and annual reporting processes to be more long-term focused. Now, let's look at a more general approach to the issue of corporate governance, one that supports a fundamental shift in how companies work, so attitudes and practices might line up with planetary needs.

A few years ago Pavan Sukhdev, a former managing director of Deutsche Bank, wrote a book called *Corporation 2020*, in which he described what it takes for a company to start taking responsibility to serve the whole, instead of serving only the interests of its shareholders. This is a great book for anyone who wants to dig into this question more seriously, but its message is summarized in the mission statement for the Corporation 2020 initiative, which grew out of that book: [17]

> **About Corporation 20/20**
>
> What is the core purpose of the corporation? How should it be designed to seamlessly blend sustainability into its design, ownership, governance, strategy and practices? Corporation 20/20 is an international, multi-stakeholder initiative that seeks to answer these questions. Its goal is to develop and disseminate a vision, pathway for the 21st century corporation in which social purpose moves from the periphery to the heart of the organization. Such transformation is indispensable to a Great Transition toward a just and livable world.

New Principles for Corporate Design

1. The purpose of the corporation is to harness private interests to serve the public interest.
2. Corporations shall accrue fair returns for shareholders, but not at the expense of the legitimate interests of other stakeholders.
3. Corporations shall operate sustainably, meeting the needs of the present generation without compromising the ability of future generations to meet their needs.
4. Corporations shall distribute their wealth equitably among those who contribute to its creation.
5. Corporations shall be governed in a manner that is participatory, transparent, ethical, and accountable.
6. Corporations shall not infringe on the right of natural persons to govern themselves, nor infringe on other universal human rights.

These new design principles may seem like lofty ambitions, but if you look around the world, you will see that the number of companies who embrace them in some way is growing. Sukhdev was also awarded the Gothenburg Award in Leadership for Sustainable Development (just as Paul Polman was), in recognition of the impact that his ideas have had on other leaders. Why are ideas like this, focused on a new and more sustainable way of thinking about corporate governance, taking root?

Let's call it a slow awakening to the truth—the truth of our interconnectedness. Sukhdev describes how the success of a corporation, for example, is dependent on the fact that many of the actual costs and risks connected to its production of goods and services are actually covered by society, not the company. This has obviously been true for many environmental costs, such as pollution, but also for social and economic costs, such as health-care expenses or shortened life expectancy caused by fast food and tobacco. A prime example is the way public money was used to bail out financial institutions after the financial crisis of 2008 and following years—even though the crisis itself was triggered by those institutions. The companies created business opportunities, and generated profits on behalf of their shareholders, but it turns out that the risks were buffered by the public.

Part of what's happening lately is that the public has woken up to this reality and is acting, through government and civil society, to put pressure for change on corporations. But corporations are also changing from within. And as in the case of Paul Polman, they are changing because their leaders have recognized that this old way of running companies—where so many risks and costs, called "externalities," are left to be borne by the public and the planet—cannot work in the long term.

Whether because of external pressure or internal enlightenment, more and more corporate leaders are becoming change agents for sustainability. They are changing production processes, business models, and even the nature of their relationship with investors and consumers.

How can we encourage the development of more, and better, leaders for change?

Models of Leadership that Integrate Change

Peter Hawkins has developed a change approach in which he focuses on three dimensions: Strategy, Culture, and Leadership. Most real strategic changes in an organization require a shift in the underlying organizational culture. To provide that shift, leadership has to change people's behavior and often their attitudes.

To support your strategy to create a shift in a culture (and so as not to be "eaten up for breakfast" by the existing culture), leadership is a useful ingredient. Earlier in this book, we looked at another model developed by Peter Hawkins called API—Authority, Presence, and Impact. The API model is used to support leaders in changing their personal behavior, so they can more effectively create an impact in a system.

But now we have to look at a more fundamental question. How do we develop leaders capable of creating the shift toward sustainability? We need new leaders who are capable of making the strategic shift toward such initiatives as Corporation 20/20. These initiatives will succeed only if a new culture emerges—not only in organizational terms but in terms of how we do business.

Brain science tells us that our brains are quite flexible in creating new connections between neurons that build on our experiences. If we have similar experiences again and again, the brain creates more solid connections between the neurons related to those experiences, which in turn are the basis of our habitual behavior. Habits are such a huge part of who and what we are that we scarcely notice them—and that's exactly their purpose. When we have habits, we can let them run on automatic pilot and spend our conscious thinking time on more important things. And we have habits for everything: how we select and weigh information, our emotional responses, mental models and paradigms, and, of course, actions, as well as how we come up with reasons for actions.

Every little girl and boy who is born into a specific society absorbs the experience that their society provides. As they grow up in our society, for example, they notice how much people value success and material wealth. Starting already in school, there is competition to be the best in class. Sometimes a little cheating is even widely accepted, if you can get away with it—for example, fouls in sports that are not called. Winning might even require such behavior. Taking what you get, and what you can get away with, is generally rewarded. Experiences like this accumulate in the brains of every girl and boy, creating a representation or "imprint" of this outside world on their neural systems. This becomes the base of their worldview, and the culture that we all live in and perpetuate.

All leaders that emerge in a culture, through its processes of upbringing, have internalized patterns like these. So leaders for sustainability need to create the ability to grow *beyond* these cultural patterns, which are also part of themselves, embedded in the very structure of their brains. To change reality, they really have to change the way they think—and the way others think.

We need to cultivate leaders who have matured to a new way of seeing, valuing, understanding, and acting, leaders who have grown beyond the standard way of perceiving the world—as a race to see who can accumulate the most wealth—as well as our human role in creating that world.

Now our café conversation is about to become philosophical. We're going to talk about consciousness. But let's remember that changing the world has to involve changing consciousness, because how we think determines how we act to make our world, whether we follow old habits, or learn how to create new ones.

Consciousness Evolves

These days neuroscientists still debate (heatedly) whether consciousness can be reduced to neural activity or whether neuroscience can even explain anything about consciousness. The experience of being self-aware, experiencing and acting in the world, remains deeply mysterious.

But most would agree that this special experience we call consciousness was probably much less developed for *Homo erectus* than it is now for the much larger-brained *Homo sapiens*. And few would dispute that we have become more and more aware of our world and how it works. Finally, even a quick glance at the history of philosophy and religions suggests that the nature of our awareness, and especially our *self*-awareness, is changing over time.

For many years, and long before neuroscientists got involved, philosophers have speculated about this development of consciousness. And many have concluded that humanity is engaged in what we might call a process of maturation: we are getting more and more aware of our surroundings and more and more capable of interacting with them intelligently, and even (we hope) wisely. The evolution of this capacity has even been tied to our ability to adapt to a changing environment and climate, such as the arrival and departure of ice ages and the massive changes they brought to the lives of our evolutionary forebears.

Some of the philosophers who have tackled this question of the evolution of consciousness include Jean Gebser, Pierre Teilhard de Chardin, and Sri Aurobindo, who almost simultaneously developed a theory of the evolution of consciousness in the 1940s, working in different parts of the world. In the 1980s some of these theories were picked up, adapted, popularized, and spread further by many other thinkers, some of them working academically, others just acting through the popular culture.

In terms that are quite similar to the seven ways of seeing change that we described in chapter 3, these thinkers describe different kinds of mental models, frames of reference, action logics, and paradigms that structure the way we perceive the world, and ourselves in the world, and that also frame how we interact with the world. In evolutionary terms, these paradigms develop, and they provide us with an increasing ability to deal with more and more complexity.

Some of these evolving paradigms lead us humans to feeling increasingly independent from, and even superior to, the rest of the world—or even all of nature. In the earlier stages of our evolution, humans were simply reactive and dependent on what the world and nature provided to them. Sometime during the past one or two hundred thousand years, we acquired the ability to hunt, migrate, command fire, and eventually domesticate some of the animals around us. About ten thousand years ago, we learned how to manipulate nature well enough to create farms. We even began to manipulate natural systems through plant and animal breeding and reshaping the natural landscapes around us.

All of this helped humans to increase our survival, and our numbers, and improve the material lives of our growing families, clans, tribes, and eventually nations. These basic facts of human evolutionary history reflect enormous changes in our consciousness as well—changes that continue today. By learning to manipulate nature, we also began to distance ourselves from it and to see ourselves as somehow separate from it. This tendency has only accelerated in modern times, with the rise of cheap energy, sophisticated technology, and our ever-more-efficient processes of turning natural resources into the tools and gadgets we use every day. (We are, after all, writing this book on computer screens, and sending e-mails to each other on our smartphones.)

But these increasing abilities are still tied to this growing sense of separateness from nature, as well as old human-cultural habits of putting oneself, one's family, and one's tribe or nation above all others. To survive, humans got smarter—but they also became (at times) harsher and more cruel. Cheating, stealing, killing, and going to war are still very much considered a reasonable strategy for improving the well-being of one's clan or nation by a large percentage of humanity, as any newspaper can quickly illustrate.

This paradigm or way of thinking—that we are separate from nature, that we should maximize our share of its resources, and that we should compete (even to the death) with anyone who stands in our way—was somehow working, in pure evolutionary survival terms, as long as our human impact could be borne by the seemingly endless earth. We could chop trees, dig mines, emit wastes, and wage war without actually threatening the survival of our own species.

Obviously, our situation has changed. Unfortunately, this way of thinking—and the behavior that goes along with it—are leading us down an increasingly dangerous path. We now have the ability to put even our own evolution at risk. Nuclear waste will cause trouble for many thousands of years. A nuclear war could basically end civilization as we know it and render the earth virtually uninhabitable for creatures like us. And this is before we get to the controversial and pressing issues of climate change, genetic engineering, population growth, resource scarcity, and the like.

"We can't solve problems by using the same kind of thinking we used when we created them." This famous quote from Albert Einstein underlines the necessity for us human beings to grow beyond our current paradigm, action logic, frames of reference, mental models—our current consciousness. The way we think evolved naturally. Now, ironically, that "natural" way of thinking has led us into a true sustainability crisis. We have to take charge, not just of our own physical development on planet Earth and the care of its physical ecosystems; we also have to take charge of the evolution of our own consciousness.

New Consciousness for New Leaders

We believe that the challenge of leadership for sustainability is not just about changing a company's production process, or leading an energy program. Ultimately, leaders have a responsibility to lead a transformation in the way humanity thinks. That responsibility suggests that we also need to think about leadership itself—what it is, and how we pursue it.

One of our favorite models for thinking about leadership comes from David Rooke and William Torbert, two professors of management, in their article "Seven Transformations of Leadership" in the *Harvard Business Review* (chosen as one of the magazine's ten best leadership articles.)[18]

In the article Rooke and Torbert explore different "action logics" (mental models) that drive the understanding and behavior of leaders. The model has since been specifically adapted to sustainability leadership, notably by Barrett Chapman Brown.[19]

Rooke and Torbert differentiated seven stages of what we might call the evolution of leadership, based on extensive surveys conducted around the world over twenty-five years. In the list below you will see the seven categories, ranging in order from the most "advanced" stage to the least, and how leaders tended to fall into them.

We'll explain these concepts in a moment. First, we can tell you that research was conducted about the efficiency of those different types of leaders in change processes. Opportunists, Diplomats, and Experts—who together represent 55 percent of the whole population of leaders surveyed—were significantly less effective

Seven Stages of Leadership	
Action Logic	Percentage of Sample Profile
Alchemist	1%
Strategist	4%
Individualist	10%
Achiever	30%
Expert	38%
Diplomat	12%
Opportunist	5%

than the Achievers. When it comes to transforming an organization or dealing with a complex and innovative change process, the Individualists, Strategists, and Alchemists—who represent the top three tiers and roughly 15 percent of the overall leadership community—showed the greatest capacity and ability to handle those changes.

So let's have a look at these seven levels of leadership. What are they?

The Opportunist: The Opportunist is mainly driven by personal needs and wishes. He or she is striving to win and get the best out of a situation, for himself or herself. To achieve these personal goals, the Opportunist is willing to use and manipulate situations, as well as other people.

The Diplomat: For the diplomat, it is important to belong to, and feel part of, a group. The Diplomat's behavior is driven by the desire to achieve that goal. She or he will tend to avoid conflicts and will be on the lookout for both obvious and hidden rules, in order to live by them.

The Expert: Knowing a topic well is the terrain of the Expert. He or she is used to looking at facts and figures. For the Expert, reality and the world need to be represented in figures and hard data. The Expert is happiest using knowledge and expertise to drive efficiency.

The Achiever: A "stretch goal" is the fuel that achievers live by. They are used to working hard to achieve their targets and produce deliverables. With their passion for accomplishment, they often create a very encouraging working environment.

Rooke and Torbert describe these first four levels of leadership as those that build on a conventional action logic, which functions best if the world is a stable and predictable entity. But they don't function as well in a world that is ruled by complexity and systemic interactions—which is the world of sustainability. The last three levels are based on a postconventional action logic; that is, a new paradigm.

The Individualist: At this level of consciousness, the leader has the ability to use *different* action logics, in different contexts. The Individualist has the ability to navigate different systems, to communicate and interact with people from different backgrounds and action logics. This helps him or her deal efficiently with conflicting emotions and diverse opinions. The Individualist is open to listening fully, notices underlying patterns, and also recognizes inner or hidden habits that might be limiting his or her effectiveness.

The Strategist: The strategist has the ability to focus on long-term developmental processes. She or he knows that these journeys are full of unpredictability and that it's important to have some general principles to guide you through. To encourage learning and development, the Strategist uses dialogue structures and double-loop learning. The aim of the Strategist is to create learning systems that can be self-sustaining.

The Alchemist: The alchemist often is a charismatic personality who lives according to very high moral standards. She or he is connected to a sense of truth as such or is trying to find truth in every moment. (Gandhi, a classic Alchemist, titled his autobiography *The Story of My Experiments with Truth*.) Through this continual search for truth, the Alchemist is often able to light an inner fire in others and give them a sense of passion and a striving for something bigger.

Barrett Brown added and described an additional level, "The Ironist," which he found in the earlier work of Torbert and Susanne Cook-Greuter, referring to those who can transcend and distance themselves from specific paradigms (in the way that irony creates a sense of distance) and create the space for new paradigms to emerge.

Brown conducted a study to analyze these three later stages of action logic in relation to leaders for sustainability. The underlying assumption in his study was that a sustainability change process is dealing with higher levels of complexity than any other

Action Logic	Main Focus	Characteristics	Strengths as Organizational Member	% of US Adult Population (n = 4,510)
Ironist [under research]	Being; experience moment to moment arising of consciousness	*Institutionalizes developmental processes through "liberating disciplines."* Holds cosmic or universal perspective; visionary	Creates the conditions for deep development of individuals and collectives	0.5%
Alchemist	Interplay of action, awareness, thought, and effects; transforming self and others	*Generates social transformations.* Integrates material, spiritual, and societal transformation	Good at leading society-wide transformations	1.5%
Strategist	Linking theory and principles with practice; dynamic systems interactions	*Generates organizational and personal transformations.* Exercises the power of mutual inquiry, vigilance, and vulnerability for both the short and long term	Effective as a transformational leader	4.9%

Advanced Leadership Levels. The three most advanced levels of leadership, according to Barrett Brown. Table courtesy of Barrett Brown.

change process (even though many of these are also complicated and complex) because it includes social, ecological, economic, and well-being issues. In sustainability transformations not only do you have more and different information to process but also a very high number of the issues that need to be addressed are interrelated. One has to look not only at the direct cause-and-effect interrelations but also at the interrelation across time and space. This really raises the bar on what a leader needs to be able to deal with!

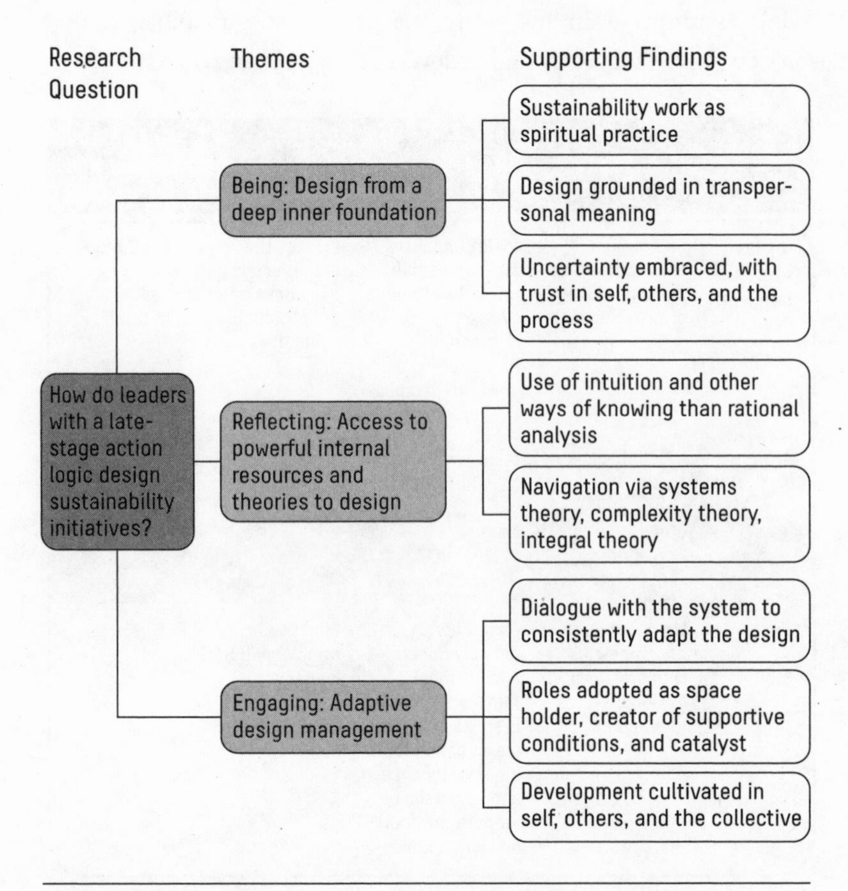

How Late-Stage Leaders Approach Sustainability Initiatives. Barrett Brown's conception of how late-stage leaders design their approaches to action. Illustration courtesy of Barrett Brown.

Consider the fact that some of the ecological problems people are facing are the result of actions taken far away. For example, the acid rain that polluted lakes in Norway in the 1980s was put into the atmosphere through very high chimneys in the Ruhr Valley of Germany. In those days people believed that to keep the air pollution lower in the neighboring areas around the industries that burned a lot of coal it was a good idea to build much higher chimneys. But one unexpected effect of building such chimneys was that in 1980 there was six times more sulphur coming down on Norway than the Norwegians themselves were putting into the atmosphere.[20]

In Brown's study. he interviewed and assessed thirty-two sustainability leaders and change agents. With thirteen of those, he conducted in-depth analyses. According to his findings, among the thirteen leaders and change agents there were six Strategists, three Alchemists, and two Ironists.

All of them were coming from mid- and senior-level management in business, government, civil society, and consultancy, and each had been engaged in initiatives that impacted more a thousand people at a time.

The leaders and change agents had been doing sustainability work for at least the last three years, and they were coming from North and South America, West Europe, and Oceania.

The thirteen comprised five women and eight men, with ages spanning between thirty-three and sixty-six.

Brown identified three different themes the thirteen leaders and change agents had in common, with some specific differentiation of those themes:

Theme 1—Being: Design from a deep inner foundation. These thirteen leaders saw their work as a spiritual practice, grounded in a sense of deep meaning. They embraced uncertainty and felt a high sense of trust in themselves, others, and the process itself.

Theme 2—Reflecting: Access to powerful internal resources and theories. The group expressed comfort with the use of intuition and other ways of knowing not limited to rational analysis. They navigated with the aid of systems theory, complexity theory, and other relatively advanced ways of understanding their reality.

Theme 3 – Engaging: Adaptive design management. The leaders engaged in continuous dialogue with the system to consistently adapt their design. They cultivated a sense of development in self, others, and the collective whole. And they adopted the role of being a catalyst, a creator of supportive conditions under which change could occur.

From Brown's study of these top sustainability leaders, we can conclude that maturing along the different levels of action logic brings new capacities and capabilities for the individual, giving them a different perspective, even a different perception of reality, and an altered way of understanding and reasoning. With leadership, as in any area of learning, as you mature to more advanced levels you still have access to those things that you learned earlier and can use them.

But you probably can't conceive of the options available to you at the advanced levels until you actually get there.

A ten-year-old child might be able to easily multiply numbers up to twenty. But it's very difficult for him to understand what percentages mean or differential equations. But when the moment arrives that he has moved into understanding percentages or differential equations, he still can multiply numbers up to twenty.

If you're somebody who has matured to a higher level of action logic, as described in these three categories of Strategist, Alchemist, and Ironist, you still have the ability to act like an Expert or an Achiever when you need to (which can be very useful). *Advancing to each new level gives you a higher degree of freedom to approach issues with different options.*

On the other hand, if you discover that your level of action logic is not so advanced—and needs to be *more* advanced—then you'll have to do some work so as not to remain stuck with a less flexible, more mechanical understanding.

Barrett summarized the characteristics of the higher action logics as involving the capacity to consistently work with the following:[21]

- Having a systems capacity and a "unitive" perspective on reality
- Simultaneously holding in mind and managing conflicting frames of reference, perspectives, and emotions
- A profound access to intuition, where the rational mind is seen as one tool and not the only or the dominant vehicle to understanding
- A deep acceptance of self, others, and the moment without judgment and an ability to be open, flexible, attuned to what is happening in the moment
- Toleration of ambiguity and the ability for collaborative engagement, within that ambiguity, to cocreate something new
- Frequent experiences of the "flow" state in consciousness

Evolving as a Leader

Every evolution, whether in the physical or the mental sphere, is a kind of dialogue and interaction process between the entity that is evolving and the surrounding ecosystem. Changes in one produce the impulse of adaptation in the other. In that spirit we would like to conclude this chapter with a quote from leadership researchers K. M. Eigel and K. W. Kuhnert:

> The future of our organizations depends on successfully identifying and developing all leaders to higher [developmental levels]—to a place of greater *authenticity*—so that they can respond effectively to the increasingly complex demands of our times.[22] [We added the italics for emphasis.]

Being a change agent for sustainability seems not to be a task for everybody. At the same time, the engagement with sustainability seems to initiate and support both our professional and personal growth. In our final chapter we would like to focus on some intentional steps you can take to support that process.

16

Growing into Possibility

KEY MESSAGES

- You can get help for dealing with challenging situations by linking up with other people for supervision, "intervision," or group support processes.
- You can sustain your own emotional energy as a change agent by learning to shift from empathy to compassion—while retaining the capacity for empathy.
- Here are some simple practices for expanding your sense of what "I am" or "i AM" means, to include more of the people and world around you. If you use these practices, they can be very powerful tools for improving your effectiveness—and your quality of life.

This will be our last cup of coffee here in our virtual café, so we'd like to focus on the most important part of any change process: you.

Abraham Lincoln once said, "Give me six hours to chop down a tree, and I will spend the first four sharpening the axe."

As a change agent you will face a lot of challenges on your way to achieving your goals; it will be crucial to check your tools and sharpen them regularly and diligently. And the most valuable tool you have is you, yourself.

On this last part of our little journey, we will consider the different aspects that need to be sharpened regularly, starting with your own continuing process of professional development and how you find support for dealing with difficult cases.

Supervision and Intervision

A standard of professional consultation to get this kind of support has been established within the helping professions, such as psychotherapy, social work, and to some degree also among medical care. That standard is called *supervision* (when you are conferring with someone senior to yourself), or *intervision* (when you are getting peer-to-peer support). In the medical field so-called *Balint Groups*—where doctors gather to talk through difficult cases—are also popular. The intent of these approaches is to help the psychotherapist, social worker, or medical doctor reflect on what he is doing with his patients or clients, how he feels in specific situations, what is happening at the more subtle relationship level between him and the patient or client.

We can also see psychotherapists, social workers, and medical doctors as change agents focusing mainly on helping the individual to change. As our example in chapter 11 about the study of bypass surgery patients showed, the path to success is often not on the level of technical expertise. Whether the patients change their lifestyles to avoid another bypass surgery depends on how much impact the medical doctor creates with regard to the patients' behavior.

These professional change agents have learned that to create and maintain a high level of impact, we need to constantly look in the mirror. We need to find out how we are actually doing things and how we can do them better. "Better" often means reflecting on how we, as change agents, might be the limiting factor of change.

Systems theory teaches us change agents that we must pay attention to the very subtle interactions and the overall relationship between us as change agents and the systems we intend to change. When the system doesn't change according to plan, we need to take a moment and reflect on our assumptions, our general approach, our actions, our relationships to the client system, and other factors. We need to constantly remember that if one part of a system changes, the whole system needs to adapt to that change. We can't escape the fact that the system reacts to what we do, whether we continue to do the same old thing or try something very new. Here's an example.

Axel: I was once engaged to work with an organizational change process that was initiated by the CEO of that organization, and she really wanted to be involved in it. My role was to work with the whole leadership on the three top levels. I noticed pretty early on that the relationship between the head of the organization and everyone else was not good at all. But as everybody kept giving positive "lip service" to the change process, I didn't pay too much attention to that, which at the end turned out to be a big mistake.

Part of the process design was to work in small groups on specific issues. The atmosphere in some of these groups was just terrible. Participants were constantly complaining about the whole organization and about the other members of the leadership team. They were questioning the whole change approach, and they were subtly criticizing me personally. I attempted to stay in my role of professional change agent and listen to their concerns, trying to understand what was behind the issues and to shift their attention back to the task we planned to do.

After each of these sessions I felt really exhausted and frustrated. When I talked about it in my intervision group, I realized a couple of things.

- *I really needed this assignment, as my pipeline of work was pretty limited in those days. This personal situation of need triggered in me an attitude of not being prepared to confront the people involved. I somehow assumed that being too confrontational would cause the managers to kick me out of the process, so I decided to address the issues in a diplomatic way.*

- *I was avoiding addressing the conflict on the top level because I was hoping that through some progress in the change process, the leadership problems would be resolved.*
- *In the small group processes I jumped too fast back to the working level, without really realizing the frustration among the staff. At the same time, I didn't stand up enough for myself and set clear communication rules.*
- *Last but not least, I had not created a good and trustful relationship with the members of those groups.*

When I finally understood all of that, I realized that I personally had to shift. In my professional approach I needed to make a clear distinction between my needs—such as my need for making enough money—and the needs of the organization in this change process.

As a change agent I needed to confront the organization much more. So I went back and had some good conversations with the top-level team about their underlying conflict. Although I had avoided this earlier, I realized afterward that everybody was relieved that somebody (me) had finally stood up and spoken plainly about the problems they were facing.

At the next working group meeting, I started by giving them clear feedback about their behavior and communication style and told them that it was not my role to push them into any change but that I could support them in their process. I asked them to discuss two questions for thirty minutes:

- *Do you want to work with me actively in the change process?*
- *If yes, what are the three top issues you feel are most important to focus on in relation to the overall goal?*

Then I left the room for half an hour. When I came back, we started a good and open conversation, which was the start of an effective collaboration.

Without the support from my intervision group, I would probably have struggled much, much longer and maybe never reached clarity about how I was an important part of what was limiting the success of the process. Without that insight we might have never reached our targets.

Here is our advice: Take your personal learning and reflection very seriously. See the great resource that you carry with yourself all the time—you and your ability to reflect on your own patterns. If you are trapped in your patterns and habits, you will definitely limit or lose your ability to impact an organization, a change team, or an initiative.

Getting some coaching for yourself, or connecting up with professional supervision and intervision, is not only a great way to make your life easier. It can also help you increase those impacts you want to create.

For your own coaching and supervision, take a look around and search for people with experience in the field where you want to develop further. Ideally, it is good to find somebody in your geographic area, but we also have had very good experiences doing coaching and supervision virtually, using teleconferencing technology.

Intervision is even easier. You just need a couple of colleagues to meet with, once a month or every two months for three hours. You can also use the Peer Consulting Process that we described in chapter 12.

If you want to develop and find support for self-improvement, the important thing is not just to think about it but to *do* something about it.

From Empathy to Compassion

Many colleagues in the field of sustainability change agentry are very committed and passionate about what they do. Observing the current reality of endangered species, soils soaked with oil, huge amounts of plastic floating in the oceans, starving children in a world of plenty, and so much more triggers strong feelings. Many of us feel pushed to try to change the situation, to make things better. On the other hand, we also are faced with the slow pace of change and the threat that we're not going to make enough of a difference in time to prevent even more serious environmental damage or human injustice.

As a sustainability professional you will often be challenged by the reality of what you are working on and the very mixed and even conflicting feelings that come along with it, including deep sadness, frustration, anger, hope, fulfillment, happiness, and love. For most people managing these emotions becomes an integral part of the work itself. We have seen many professionals for whom a deep engagement with sustainability and change created the greatest sense of meaning in their lives. But we have also seen those who became burned out and who ended up cynics.

As a sustainability change agent, you should be aware that this is not a short-term mission. In all likelihood this work will not be completed in your personal life span. You are a part of a movement. There is every possibility that we humans will manage to use our collective intelligence and wisdom well enough to find a sustainable balance between our well-being and needs (in fairness and justice to all) and the well-being and needs of all other species and the rest of nature. Maybe we will even find a way not only to minimize our negative ecological footprint but to create a positive ecological handprint, in which our economy acts to restore and protect nature instead of degrading it.

But on the way to that visionary destination, you need to take care of your own well-being and make sure that you don't suffer from any negative side effects of making the journey . . . which brings us to a very interesting finding in brain research about the difference between empathy and compassion.

In our brains we have some specific cells called "mirror neurons." These cells are activated (among other ways) when we observe the emotional states of other people in a very specific way—a mirroring way. If you see that somebody is sad, your mirror neurons activate sadness in your own emotional system. The same is true for despair and anger—but also for happiness and delight. In one experiment people were asked to watch short films in which they observed other people suffering, and then their neurohormonal reactions were tested. Just watching people suffering created a stress reaction in the experimental subjects that was almost as strong as if the subjects were suffering physically themselves. Reacting with empathy to all the negative indicators of an unsustainable way of life that we see around us might induce stress to your body and, ironically, make you as a change agent less sustainable.

But this neuroscience research also discovered something else, which is very interesting and liberating. The researchers created an additional experimental group and taught these subjects a specific Buddhist meditation technique designed to develop *compassion*. In this meditation you first relax and become mindful (attentive), then you create a feeling of love and compassion inside yourself. In the next step you try to expand that feeling outward, first to yourself, then to the people you love, and then to other people and beings who are suffering. Finally, you even extend the feeling of love and compassion out to those people who are committing suffering to others. When this group of experimental subjects had learned and practiced this "Metta meditation" of loving kindness for some time, they participated in the same experiment, watching similar video clips that showed people suffering. This time their neurohormonal reactions were different. The participants were internally and subjectively experiencing a feeling of love and compassion, instead of suffering, and the biological indicators also were those associated with a positive internal state. Performing the mental exercise of developing loving kindness not only reduced the physical stress symptoms associated with observing suffering but also increased positive emotional states.

It is important to underscore here that the participants were not being encouraged to avoid their feelings when observing suffering but rather to *train* their feelings, shifting from empathy (feeling what the other person feels) to compassion (feeling love for that person and a desire to help). Empathy is a very useful and necessary emotional skill, but compassion is an emotion that can sustain us in the work of creating sustainability.

These are tools and methods you can use to "sharpen your axe" in subtle but very real ways. These methods can help you stay fit and remain in the game for the long haul.

Staying fit is not just a question of physical fitness. Of course we would suggest that you treat yourself well, and make sure to maintain a good diet and enough time and space for physical exercise. Any friend in a café would tell you that! But what is also particularly important, in your work as a sustainability change agent, is the care of your mind and your emotions.

Moving from "I am" to "i AM"

You may have noticed a few references to Buddhist thinking and practice in this book, and that's no accident. Both of us have practiced Buddhist meditation (different kinds, at different periods of our lives). We believe that some of the tools and approaches from Buddhist tradition are universal and not linked to any religion: you don't have to be a Buddhist to use them. And we have both noticed remarkable similarities between systems thinking and Buddhist philosophy.

One of the most obvious similarities is the emphasis on the concept of interdependence. Here are two illustrative quotes:

- "One of the most important concepts in Systems Theory is the notion of interdependence between systems (or subsystems). Systems rarely exist in isolation. . . . It is important for an analyst to identify these interdependences early. It may be the case that changes you make to one system will affect another in ways you haven't considered, or vice versa." (From a university curriculum on systems theory for computer science, at the University of New Brunswick)

- "Thus interdependence is a fundamental law of nature. Not only higher forms of life but also many of the smallest insects are social beings who, without any religion, law, or education, survive by mutual cooperation based on an innate recognition of their interconnectedness. The most subtle level of material phenomena is also governed by interdependence. All phenomena, from the planet we inhabit to the oceans, clouds, forests, and flowers that surround us, arise in dependence upon subtle patterns of energy. Without their proper interaction, they dissolve and decay." (From the Dalai Lama, in his book *The Compassionate Life*)

You can see that this notion of interdependence is at the heart of compassion, which is at the heart of sustaining our action to create a sustainable world. As the Dalai Lama also writes, "In today's world, every nation is heavily interdependent, interconnected. Under these circumstances, destroying your enemy—your neighbor—means destroying yourself in the long run. You need your neighbor."[23]

The depth of these interconnections can be illustrated by thinking about a systems diagram. When making such a diagram, you try to identify all the important cause-and-effect relationships among a system's parts and subsystems. While this is framed as an intellectual exercise, it very often requires a degree of compassion to do it right: who and what is being affected here? Are there any negative impacts—including any suffering, in nature or people—that we have not yet thought about?

Also, when you look at any part closely enough, you can see that it is, itself, a system. By thinking about interdependence, the links *inside* a subsystem, you can discover more possible leverage points for positive change. Here is a simple, but typical, example.

Alan: Once, pretty early in my consulting career, I was working with a small international organization on its process of stra-

tegic development. We were going to run a workshop to develop a new vision and strategy, so I was interviewing people about that in advance of the planning session. I was very focused on the task I was given: examining the organization's mission, its core competencies, how its different parts worked together, and how it interacted with stakeholders and donors. I thought I had it all pretty well mapped out. But something just felt wrong or inadequate about my analysis. To be honest, I was having trouble seeing anything that could be substantially improved. And then I discovered, after a couple of weeks of work, that the organization had already been down this same road before, with two previous consultants. Yet it was still having the same difficulties when it came to achieving good performance.

So I literally just sat with this problem, in Buddhist meditation fashion. I let my inner eyes just roam around inside my little mental map. And I realized that I had been thinking of the management team as one "thing" in that strategic process. I had not looked inside that team, to understand its dynamics.

Of course, once I started looking and asking questions, I discovered that this little management team was a very complex system all by itself. And it was inside that "subsystem"—specifically in relationship to how the leader was communicating with his deputies (who were actually suffering, in ways I had not noticed before)—that I finally began to see where the problems, and the leverage points, were to be found.

In Buddhist practice you become more aware of interdependencies by sharpening your awareness and looking more deeply into reality. When you do that, you find that nothing exists just by itself. Everything consists of relationships with other things, other aspects. A feeling that you are having might have been triggered by an action taken by somebody else, who in turn was motivated by a certain set of circumstances. Even a table, which appears so solid and permanent, was not a table fifteen years ago: it was a tree. Without the ideas of the designers, the work of the person who cut the tree, and the carpenter with her tools, the table would not exist.

In our day-to-day experience we tend to create a perception of the world around us as solid, with very simple cause-and-effect mechanisms, even though the reality is much more complex and interconnected. This day-to-day perception starts with the solid experience of "I," which is me. Everything else is not "Not I," or "you," or the "outside."

In our industrialized world, most of our psychological, educational, cultural, and economical development processes are geared to creating strong "I's" that can compete with other "I's." The fittest "I's" go the farthest. As long as we build reality *only* on that concept, it will be difficult to move into a truly sustainable relationship with our world.

In 2000 Axel published a book in German called *Liebe und werde der Du bist* (Translation: "Love and become who you are") on that very topic:[24]

> *Axel: I wrote that book as the result of a personal quest, following the Buddhist path with my teacher H. H. Gyalwang Drukpa and working as a psychotherapist and coach. In the Buddhist realm unconditional love and compassion are a main vehicle to liberation. And liberation is a state of overcoming duality and entering into the state of oneness or emptiness. In psychotherapy we are often aiming to help people integrate aspects of their personality that they have placed outside the felt sense of "I" that they usually experience.*
>
> *The normal concept of "I am" implies that there is something which is "Not I" or something that "I am not." For example, you might identify yourself as a good-looking, successful, hardworking businessperson, and you don't want to accept the side of yourself that feels insecure or that might have gained some extra weight in the last years. As individuals, we often put quite a lot of energy into dealing with those aspects of ourselves that we don't accept*

as "me." Sometimes we just ignore those aspects by not looking clearly into the mirror. But sometimes our fight against these other aspects of ourselves gets stronger, and we might choose to have cosmetic surgery to get rid of our wrinkles or take tranquilizers to overcome our insecurity and anxiety.

In a successful psychotherapy or coaching process the patients or coachees often come to a state in which they start to accept what is and to integrate those aspects of themselves into their self-image. They learn to expand this concept of "I am" and start integrating more things into it, things that they previously did not accept.

This process of integrating is supported by two aspects you can find in what we call "love." The first aspect of love is embracing what is, and a second aspect is uniting what was separated. When you love, you shift into a different way of being for a moment: you go from "I am" to "i AM." The big sense of "I" loses importance and becomes a smaller "i." At the same time the being aspect—the "AM"—becomes larger and more conscious. It's in this larger, being aspect, the "AM," that we feel connectedness, unity, and oneness.

This journey toward expanding our sense of self is not limited to the social and psychological aspects of our feeling of separation. It also extends to our relationship with nature. How can we feel more connected to other people, and to nature, rather than perceiving ourselves as separated? After all, the connections are the reality!

Developing this expanded, felt sense of "i AM" is extremely helpful when working with complex systems. It sharpens the senses, to the point that we don't really have good words to describe the sensory experience. You can become very sensitive, for example, to what is happening in a group of people, to the point that people start talking about a "group field" or "the energy in the room." You might notice that your emotions or even bodily sensations start to respond differently to different situations in a group, and you have a feeling of "knowing"—without always knowing how you know—whether the group process is running smoothly or is about to get stuck. There is nothing really strange or mystical about these feelings or experiences; they are simply a result of working with your mind to expand your sensitivity to the people and things and dynamics you see around you, by including more of that into your sense of "i AM."

You can follow that concept even further when you work with people who are resistant to what you are trying to do in a change process. If you can try to integrate the force of their resistance—and especially the motivations for it—into your perception of the whole, and even into the process itself, it can have a big impact. As long as it remains "out there," it's just an irritant. But if you can embrace it, look at it from the inside, you may discover that this resistance is playing an important part in the system. Maybe it's slowing things down in a good way, to a speed that works without tearing the system apart. Or maybe the resistance is pointing out a problem that might occur later. You won't know until you look!

Of course, understandably, one part of us just doesn't like having to deal with resistance and would prefer to get rid of it. But trying to get rid of parts of the system, even irritating parts, almost never works. The more advanced step is to understand those parts more deeply and then help the system shift to a different constellation, a different level or type of complexity, where it can integrate these seemingly contrary and conflicting aspects. It's not about one side winning over the other side. You, as a change agent, need to have the maturity and skill to visualize an expanded reality—a reality in which opposing parts have become integrated and may even be the source of something new and better.

Working with your own consciousness in this way, in a change process, is very powerful, but it takes continuous practice. (We are still practicing!) So let's close this chapter—and this book—with a little exercise in focusing your mind, or meditation. This will take about fifteen minutes. You can read the exercise first, finish reading the last few paragraphs of the book—and then try it yourself.

- Sit upright and comfortably. It's often good to close your eyes, but you can also have them open and just let your seeing rest softly on a spot about five feet in front of you.
- Allow yourself to notice how you are sitting, and your breath, without any need to interfere with your breathing.
- Realize the ability of your awareness to purely notice what happens, without judging it and with a friendly attitude. Try to keep that state throughout the exercise.
- During the whole exercise there might be thoughts coming up or feelings or sensations. Just notice them in a friendly way, and go back to the actual area of focus.
- Allow yourself a couple of conscious breaths: Intentionally breathing out, you relax your body and mind a little bit more. Breathing in, you become more awake.
- Allow those parts of yourself about which you feel content and happy to come into your mind. How does "I am" feel, if you identify yourself with those aspects?
- Now allow those parts of yourself to come into your mind that you don't really like or accept. How does it feel if you are just more aware about those things that you dislike about yourself?
- Now connect to the original state of the mind, which was just watching without judgment and in the mode of being friendly with whatever is there. Try to expand that state, and embrace the aspects you don't like about yourself and the feeling of dislike itself. Make those a part of that feeling of "I am." Try to rest in this state for some time.
- Notice what happens in you, if you can allow yourself to embrace yourself more fully. Can you notice a shift from "**I** am" to "**I AM**" or "i **AM**"?
- After some time, just open your eyes or lift your vision again. Maybe look around the space where you're sitting. Do you notice anything different about how you feel, or even about what you're seeing, before and after this exercise?

This little exercise sounds easy when you read it. But doing it is much more difficult, especially if there is a strong belief and history connected to the inner dislike. But if you continue practicing the exercise, relaxing into that, there will be a moment when the shift from "**I** am" to "i **AM**" happens. This shift is the gateway to a felt perception of greater connectedness and interrelationship.

And that will help you not just in your work as a sustainability change agent but in your life.

Once you have familiarized yourself with this exercise, focusing on the internal aspects of yourself, you can learn to expand it. Start at the level of relationships. You begin with being aware of those people you like, with whom you can easily feel the sense of "WE are." Then notice those people whom you don't like, or have even stronger feelings against. When you have them in your mind, try to expand your awareness of "WE are" to include "Them." How does it feel? How does it feel when your awareness moves into the space of "we **ARE**," which is the extended version of "i **AM**"? (And again, please don't think this is a fast, easy, one-time exercise.)

On a further level you can try a similar exercise with nature. Sit in front of a tree, a lake, or a garden, for example. Let your awareness start from "i AM" and let it expand around the living things and the landscape in front of you. How does it feel if, in your awareness, the sense of separation between you and the natural world vanishes?

Over time your mind is more and more used to seeing the connections, or even the sense of oneness—that it is all part of one big system.

At that point you also might notice a fundamental change in your relation to the change processes you are involved in. You might come to a realization that these change processes are not just external engagements but more like mutual journeys, in which the things that are happening on the outside are tightly linked to internal experience. Then, in reading your own inner signals, you will get crucial information about the whole, expanded reality.

This journey never ends. It is a journey during which learning and making an impact are truly married.

Good luck!

—Axel and Alan

ACKNOWLEDGMENTS

This book would not have been possible without the contribution of many others who shared with us their ideas, methods, and experiences. We have tried our best to interpret those ideas for you, the reader. Any fault in what we presented certainly lies with us, not them, and we urge you to explore their work directly. References and links are provided in the Notes.

We cannot individually thank everyone who deserves it, for the list would be too long. But special thanks goes to Peter Hawkins and other colleagues at the Bath Consultancy Group (now part of GP Strategies Limited) for their generosity in sharing their deep understanding, methods, and tools for coaching and cultural change.

Many of the systems-related ideas and practices presented in this book were either invented or developed to maturity thanks to the annual meetings of the Balaton Group, a remarkable global network of inter-disciplinary thinkers and doers. Many thanks to all the members and especially the founders, Dennis Meadows and Donella Meadows.

We also extend heartfelt thanks to our colleagues in the AtKisson Group and Center for Sustainability Transformation, and the many graduates of our training programs, who have put this way of working with change into practice. Their use and adaptation of these ideas in many different contexts around the world has been an inspiration to us, and given us the confidence to write this book and put it out into the world.

And finally, we are very grateful to our editor at Chelsea Green, Joni Praded, and the whole team at that wonderful publishing house for believing in this book and for working so thoughtfully, diligently, and creatively to bring it to you.

But we should also extend our thanks to you, the reader. If you have come this far, it means that you have done us the great honor of investing your time in reading our book. We hope that it helps you go a bit farther down the road of transformation—and we wish you the best of luck in your own work to create a better and more sustainable world.

— Axel Klimek and Alan AtKisson

NOTES

1. Scott Keller and Carolyn Aiken, *The Inconvenient Truth about Change Management* (New York: McKinsey & Company, 2009).
2. Gervase R. Bushe, "The Appreciative Inquiry Model," in *Encyclopedia of Management Theory*, Eric H. Kessler (Thousand Oaks, Calif: Sage Publications, 2013).
3. Alan AtKisson, *Believing Cassandra*, 2nd ed. (Washington, D.C.: Routledge/Earthscan, 2010); and *The Sustainability Transformation*, (Washington, D.C.: Routledge/Earthscan, 2010) (updated paperback edition; note that the original hardback edition had a different title, *The ISIS Agreement*). You can also watch a TEDx talk about the Amoeba at http://bit.ly/AlanTED2014.
4. See www.collectiveleadership.com.
5. For more information on the Dialogic Change Model, see Petra Kuenkel, Silvine Gerlach, and Vera Frieg, *Working with Stakeholder Dialogues* (Stoughton, Wis.: Books on Demand, 2011). For more information on Kuenkel's other change methodology, the Collective Leadership Compass, see Petra Kuenkel, *The Art of Leading Collectively* (White River Junction, Vt.: Chelsea Green Publishing, 2016).
6. You can find out more about Pyramid, and all our Accelerator tools, at our website, where you can also download a free "lite" version of our tools for your own use. See http://SustainabilityTransformation.com/acceleratorlite/.
7. The Sustainability Compass is included in the Accelerator tools. It was invented in 1997 by Alan AtKisson, and

first published in AtKisson et al., "The Compass Index of Sustainability: Prototype for a Comprehensive Sustainability Information System," *Journal of Environmental Assessment Policy and Management*, Vol. 3, No. 4.

8. Donella H. Meadows, ed. Diana Wright, *Thinking in Systems* (White River Junction, Vt.: Chelsea Green Publishing, 2008).
9. C. Otto Scharmer, *Theory U* (Oakland, Calif.: Berrett-Koehler Publishers, 2009).
10. John Kounios and Mark Beeman, "The Cognitive Neuroscience of Insight," *Current Directions in Psychological Science*, Vol. 18, Number 4 (2009): 210.
11. See "Change or Die," by Alan Deutschman, *Fast Company*: http://www.fastcompany.com/52717/change-or-die. An earlier version of this piece also appeared in the May 2005 issue of *Fast Company*.
12. For a thorough exploration of coaching, see Peter Hawkins and Nick Smith, *Coaching, Mentoring and Organizational Consultancy: Supervision, Skills & Development* (London: Open University Press, 2013).
13. The origin of the GROW model is unclear, but several people were involved in developing and popularizing it, including Graham Alexander, Alan Fine, Sir John Whitmore, and Max Landsberg.
14. See Alan AtKisson, *Believing Cassandra: How to Be an Optimist in a Pessimist's World* (London: Earthscan/Routledge, 2nd Edition, 2010); *The Sustainability Transformation: Accelerating Positive Change in Challenging Times* (London: Earthscan/Routledge, 2nd Edition, 2010); and *Sustainability is for Everyone* (Hofheim, Germany: Center for Sustainability Transformation, 2013).
15. Alan often tells this story as a song, called "The Parachuting Cats." You can see him perform it in a video at http://bit.ly/AlanTED2014. For a review of the scientific evidence supporting this story, see http://catdrop.com.

16. Nathan Bennett and G. James Lemoine, "What VUCA Really Means for You," *Harvard Business Review*, January-February 2014, http://hbr.org/2014/01/what-vuca-really-means-for-you.
17. To find out more about the Corporation 2020 initiative, visit their website: http://www.corporation2020.org.
18. David Rooke and William Torbert, "Seven Transformations of Leadership," *Harvard Business Review*, April 2005, accessible here: https://hbr.org/2005/04/seven-transformations-of-leadership.
19. Barrett Brown, "Conscious Leadership for Sustainability: How leaders with a late-stage action logic design and engage in sustainability initiatives," PhD thesis, Fielding Graduate University, 2012. The whole thesis can be downloaded here: http://integralthinkers.com/wp-content/uploads/Brown_2011_Conscious-leadership-for-sustainability_Full-dissertation_v491.pdf.
20. "Da liegt was in der Luft," Der Spiegel, 23 Nov 1981, http://www.spiegel.de/spiegel/print/d-14349434.html.
21. Adapted from Brown, "Conscious Leadership" (including his citations of other sources).
21. K. M. Eigel and K.W. Kuhnert, "Authentic Leadership Theory and Practice: Origins, Effects, and Development," *Monographs in Leadership and Management*, Vol. 3 (Bingley, UK: Elsevier, 2015), 357–385.
23. Dalai Lama and Victor Chan, *The Wisdom of Forgiveness* (New York: Penguin, 2005).
24. Axel Klimek, *Liebe und werde der Du bist* (Petersberg, Germany, Via Nova, 2000).

INDEX

ABOUT THE AUTHORS

Axel Klimek is the cofounder and managing director of the Center for Sustainability Transformation. He has worked in more than twenty-five countries on four continents, and within a wide spectrum of contexts—helping leaders, organizations, and developmental programs manage complex change processes and improve their performance. His clients have included the African Union Commission, Canon Europe, EY, PWC, Allianz, GIZ, Lufthansa, Unilever, and T-Systems. He lives in Germany.

Alan AtKisson, CEO of AtKisson Group and cofounder of the Center for Sustainability Transformation, was inducted into the International Sustainability Hall of Fame in 2013. He has advised governments, cities, and organizations around the world, including Nike, Levi Strauss, Toyota, WWF, and the United Nations Secretariat. His six previous books include the Amazon bestseller *Believing Cassandra*. He is a dual citizen of the USA and Sweden, and lives in Stockholm.

For more information,
please visit the website of the
Center for Sustainability Transformation:
www.SustainabilityTransformation.com

Chelsea Green Publishing is committed to preserving ancient forests and natural resources. We elected to print this title on paper containing 100% postconsumer recycled paper, processed chlorine-free. As a result, for this printing, we have saved:

28 Trees (40' tall and 6-8" diameter)
13,200 Gallons of Wastewater
12 million BTUs Total Energy
884 Pounds of Solid Waste
2,434 Pounds of Greenhouse Gases

Chelsea Green Publishing made this paper choice because we are a member of the Green Press Initiative, a nonprofit program dedicated to supporting authors, publishers, and suppliers in their efforts to reduce their use of fiber obtained from endangered forests. For more information, visit www.greenpressinitiative.org.

Environmental impact estimates were made using the Environmental Defense Paper Calculator. For more information visit: www.papercalculator.org.

the politics and practice of sustainable living

CHELSEA GREEN PUBLISHING

Chelsea Green Publishing sees books as tools for effecting cultural change and seeks to empower citizens to participate in reclaiming our global commons and become its impassioned stewards. If you enjoyed *Parachuting Cats into Borneo*, please consider these other great books related to systems thinking and social change.

THE ART OF LEADING COLLECTIVELY
Co-Creating a Sustainable, Socially Just Future
PETRA KUENKEL
9781603586269
Hardcover • $29.95

SYSTEMS THINKING FOR SOCIAL CHANGE
A Practical Guide to Solving Complex Problems, Avoiding Unintended Consequences, and Achieving Lasting Results
DAVID PETER STROH
9781603585804
Paperback • $24.95

THINKING IN SYSTEMS
A Primer
DONELLA MEADOWS
9781603580557
Paperback • $19.95

THE SOCIAL PROFIT HANDBOOK
The Essential Guide to Setting Goals, Assessing Outcomes, and Achieving Success for Mission-Driven Organizations
DAVID GRANT
9781603586047
Paperback • $20.00

For more information or to request a catalog, visit **www.chelseagreen.com** or call toll-free **(800) 639-4099**.

the politics and practice of sustainable living

CHELSEA GREEN PUBLISHING

REINVENTING FIRE
Bold Business Solutions for the New Energy Era
AMORY B. LOVINS and ROCKY MOUNTAIN INSTITUTE
9781603585385
Paperback • $29.95

WHAT WE THINK ABOUT WHEN WE TRY NOT TO THINK ABOUT GLOBAL WARMING
Toward a New Psychology of Climate Action
PER ESPEN STOKNES
9781603585835
Paperback • $24.95

SLOW DEMOCRACY
Rediscovering Community, Bringing Decision Making Back Home
SUSAN CLARK and WODEN TEACHOUT
9781603584135
Paperback • $19.95

2052
A Global Forecast for the Next Forty Years
JORGEN RANDERS
9781603584210
Paperback • $24.95

For more information or to request a catalog, visit **www.chelseagreen.com** or call toll-free **(800) 639-4099**.